# *Wiring Made Easy: Learn the Basics of Home Wiring and Tackle DIY Electrical Projects with Confidence*

*Step-by-Step Guide for Beginners to Wire Your House and Undertake Simple Wiring Projects in the US and UK*

# Introduction

You don't always have to "call a guy" when you want to do wiring projects. Sometimes, e.g., when installing bulbs or fixing switches, you can save your money by doing the wiring yourself.

However, you just can't decide to do wiring without any basic electrical knowledge; otherwise, you will be putting your life at risk. Yes, even if you have little to no clue about wiring, you probably know that wiring is not something to joke around with as it can cost you a lot.

You probably have many questions about electrical wiring and are probably also confused about where to start. It is even tempting to give up and call an electrician but, before you do, this book will change your mind.

This book will give you all the basic information you need to know about wiring and guide you on the steps to follow to do simple wiring projects around your home, irrespective of whether you live in the US or the UK.

Among other things, you will learn about the tools you need, wiring rules and safety, mistakes to avoid, some must-know electrical terminologies you will come across, and many more. Having this knowledge will give you the courage to do wiring projects with confidence and ease.

# Table of Content

Introduction

Chapter 1: Common Wiring Terminologies

Chapter 2- Wiring Safety Precautions and Tips

Chapter 3: Simple Wiring Projects for Beginners

   1: Replacing a One-Way Switch

   2: Wiring an Electrical Socket Outlet

   3: Installing an Outdoor Lamppost

   4: Replacing a Light Fixture

   5: Installing an Electric Cooker

   6: Installing a Doorbell

   Wired doorbell

   Wireless Doorbell

   7: Installing Security System

   8: Replacing Shower Isolator Switch

Chapter 4: Common Wiring Problems and Their Solutions

Conclusion

# Chapter 1: Common Wiring Terminologies

If you don't know basic wiring terminologies, you will struggle to understand many of the concepts within this book. Let's ensure that doesn't happen by looking at commonly used wiring terms.

Some wiring terminologies you will come across are:

# Cable

Cables are sets of wires insulated using non-conductive materials and used to conduct electricity. It is very important to know the different types of cables, their properties, color codes, usage, amperage rating, and voltage rating so that you won't have a hard time starting your projects.

There are many types of wiring cables, but let's look at the few you are likely to use for house projects.

## Non-Metallic (NM) cable

NM cable, popularly known as "Romex," is the most used cable type for most house circuits like switches, appliances, outlet wiring, and lighting. One NM cable has one or more current-currying wires (hot wire), a neutral, and an earth (ground) wire which is flat in shape and covered by a sheathing or a plastic coat (usually flexible).

NM cables are made according to their sizes, i.e., gauge and amperage rating (amp). They come in different sheathing colors, and each color indicates the cable's gauge. For example:

- Black sheathing indicates 6- gauge (55- amperage circuit) and 8- gauge (40- amperage circuit) cables.
- Orange sheathing indicates 10- gauge (30- amperage circuit) wires.
- Yellow sheathing indicates 12- gauge (20- amperage circuit) conductors.
- White sheathing indicates 14 gauge (15- amperage circuit) cables.

Since these cables contain at least one hot wire, it is dangerous to work on them when there is a voltage flow.

## THHN/ THWN cables

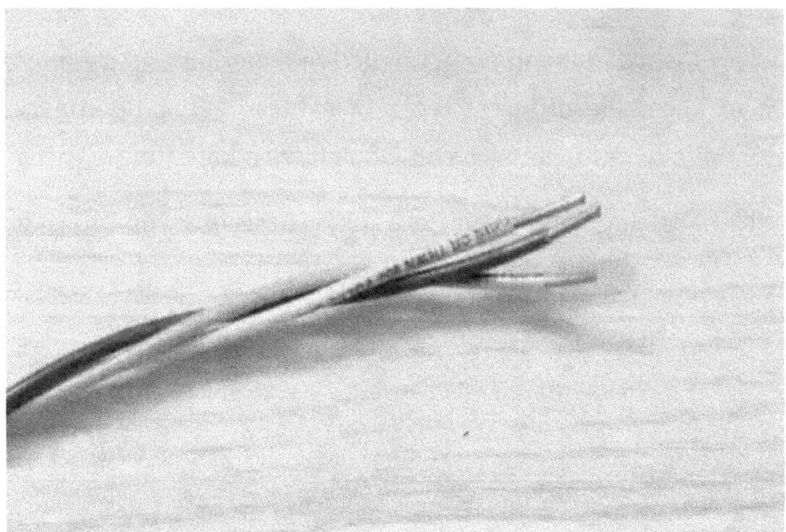

THHN/ THWN are code names for the common types of cables used in conduits (flexible or rigid plastic or metal pipes used when cables are not hidden in ceilings, ground, or walls).

These cables are single conductors, with each wire coded with colored insulation, i.e., black, orange, and red indicate hot wires, brown and white indicate neutral wires, while yellow-green and green indicate ground wires. These cable types do not have the protection of a sheath, hence the need for conduit.

The use of THHN/THWN cables is common in unfinished places such as garages or basements and short wiring connections in the house, e.g., water heater wiring.

The code names for these cables indicate wire insulation properties. T- is for thermoplastic, H- for heat-resistant, HH refers to highly heat-resistant, W- for wet areas, and N- for nylon coated.

Like NM cables, always turn off the main switch to stop voltage flow when handling THHN/THWN wires.

## Underground Feeder (UF) cables

UF is a type of NM cable that can be buried in the ground or used in any wet area, making its use common in outdoor wiring such as lampposts. It contains hot wire, a neutral wire, and a ground wire which, are insulated and sheathed with the same material (normally a grey plastic) on each wire (unlike NM, where each conductor is sheathed on a separate wrap).

UF carries a high voltage level; hence, you shouldn't handle it with a live circuit.

## Data and telephone wire

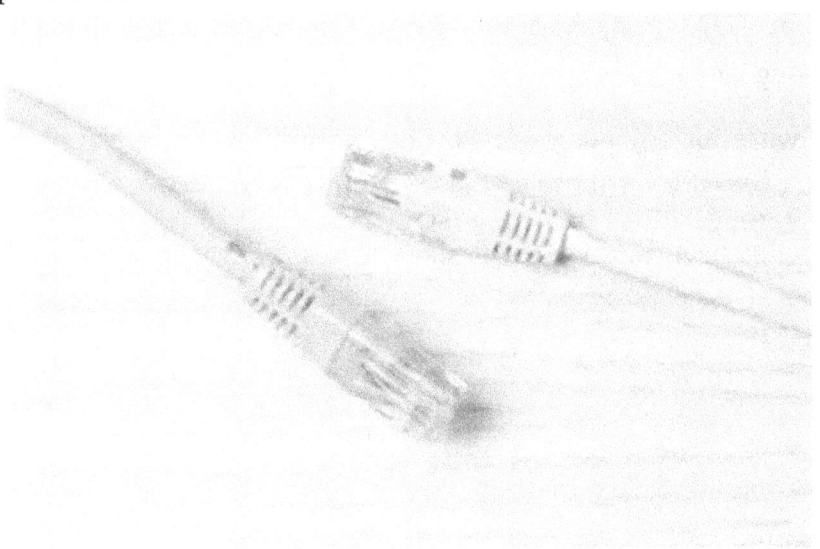

These cable types have low voltage and are used to hook-up internet and landline telephones. Most phone cables have 4 or 8 wires, with category 5 (Cat 5) being the commonly used type in households. Cat 5 has 8 wires that are wrapped together in groups of twos. Compared to standard data wires, category 5 wires are of higher quality and capability, making them usable for data transmission and phone purposes.

Although data cables are considered safe (as they carry a small amount of voltage), they are likely to come into contact with other cables, a situation that can lead to shock. To prevent this from happening, always treat all household wiring with caution and never touch bare wires unless you are sure there is no voltage flowing.

**Low- Voltage cables**

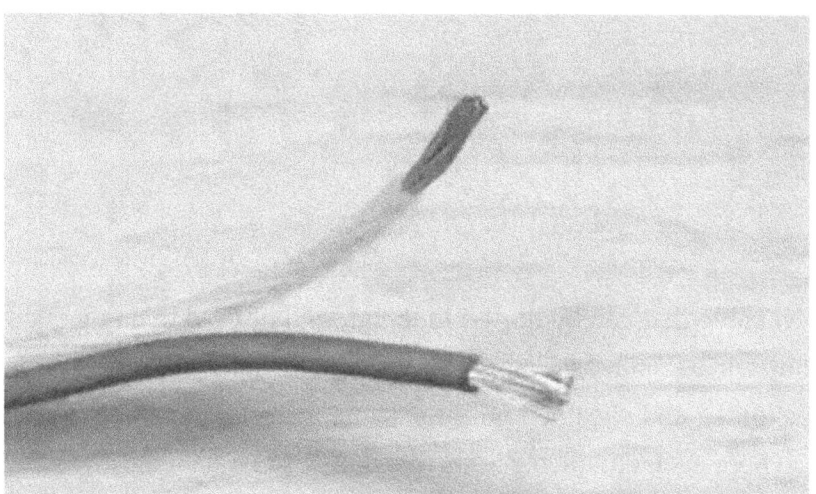

Low-voltage wires are common in circuits that require less than 50 volts, e.g., Bell wiring, speaker wiring, landscape wiring (for lighting), or sprinkler wiring. The sizes of low-voltage wiring range from 12-22 gauge. These wires are insulated and can be protected through sheathing or twisting into pairs.

Never use these wires for any purpose other than wiring low-voltage appliances like bulbs or lamps because they are very small cables and won't work on standard circuits.

These types of wires have rare electrical shock cases. However, always turn off appliances before wiring to be on the safe side.

# Switch

This refers to any device that can break or make an electrical current. When it is ON, it closes its contacts, creating a path for the flow of electrical current that allows the load, like a bulb, to receive power. When it is OFF, its contacts are open, which means current will not flow; hence, there won't be any power to consume.

Below are some easy to install switches

## Single-pole/one-way switch

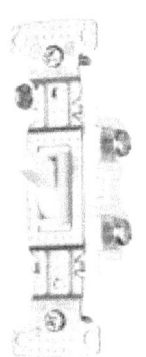

Single-pole switches are rated for 15 amps or 20 amps and are found in 2 types: Single-pole toggle switch and single-pole Decora switch. The single-pole toggle switch is an up and down, standard-type of switch that has "ON" and "OFF" marking on the toggle and is found in almost every home because it's installable in any lighting that requires up to 1800 Watts, like a Christmas tree switch, fan switch, or lamp switch.

A single-pole Decora switch, also called a rocker switch, is rectangular and larger than the toggle switch. Most of these decora switches have the "ON" and "OFF" markings in the cover plate but none on the switch.

Single-pole switches have 2 screw terminals (brass-colored) connected to hot wires, i.e., one terminal is connected to the hot wire from the source, while the other one is connected to the hot wire carrying current to the appliance. Each of these switches has ground terminals that protect the devices from excess current.

## Double-pole/two-way switch

This type of switch is installed in devices that carry around 240V; hence, its use is common in appliances that require high power, e.g., hot tub, spa, or heating system.

The Double-pole switch has 4 screw terminals connected to 2 red wires, 2 black wires, and 1 ground terminal. Two-way switches have visible "ON" and "OFF" marks and allow for the operation of more than one connection from one location.

**Three-way switch**

This type of switch is common in places with 2 entries, like the basement, garage, staircases, or hallways. They are installed in pairs and are both used to control one appliance; as a result, they do not have visible on and off marks.

A three-way switch has 3 terminals with the hot wiring carrying current from the source connecting to the terminal with "COM" (common) marking. It is also at the common terminal where the wiring carrying power to the appliance meets, forming a connection between the 2 switches. The other 2 terminals do not have the "COM" mark, making them interchangeable.

Whenever you need to install or replace a three-way switch, always ensure you get the common terminal right to avoid accidents. A three-way switch also has a ground terminal.

**Four-way switch**

Like a three-way switch, a four-way switch does not have "ON" and "OFF" marks because its main use is controlling power from more than 2 locations. e.g., living room with 3 or more entrances or long hallways.

This switch is installed between 2 three-way switches, i.e., if you want to control a light fixture or outlet from 6 points, you will have to install 2 three-way switches (one on each end) and then 4 four-way switches in the middle.

These switches are similar to the Double-pole switches, but they do not have the "ON" and "OFF" markings.

None of its 4 brass-colored terminals have the common mark on them as their main role is to act as the switching device.

## Consumer Unit

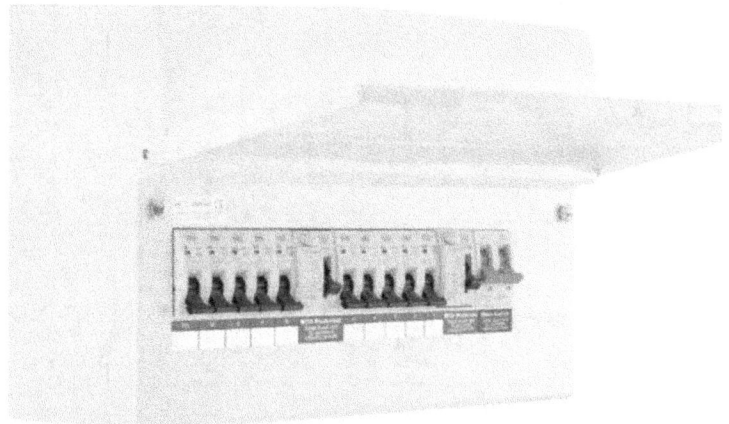

A consumer unit —also called a fuse box— controls and supplies power from the main source to different circuits in your home. A fuse box is a device that protects all appliances from damage and prevents fire outbreaks and other accidents such as shock.

Some of the components of a consumer unit are:

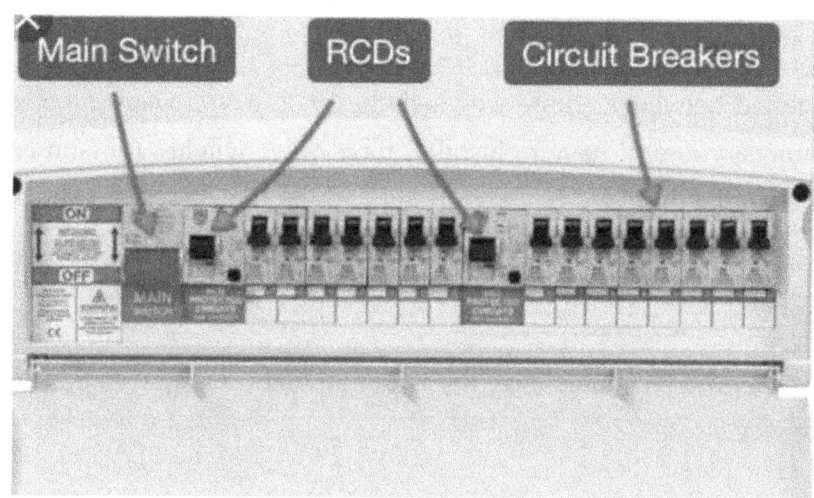

### Main switch

This is the only component in the consumer unit that's manually operable. Turning on the main switch allows power to flow from the meter to the consumer unit and different circuits, while turning it off isolates the fuse box, cutting power flow in your premises.

To avoid accidents, always turn off the main switch before doing any wiring work.

### Miniature Circuit Breakers (MCB's)

This is a small fuse-like port where all circuits are connected. MCB is automatic; hence, it turns itself off (trip) in case of:

1. Overcurrent: This is a situation where there is a defect in the circuit, e.g., when life cables are loose, or overcurrent occurs, causing MCB to trip.

2. Power overload: This happens when too many appliances are plugged into a single circuit. In such situations, the MCB automatically turns itself off.

Each home has approximately 6 circuits connected to the circuit breaker. Each of these circuits goes off any time an overload or overcurrent occurs. For example, if a 6 amps MCB detects 8 amps current flow, it will automatically trip. When this happens, you can turn the MCB back on. However, it is best to consult a professional electrician to find and fix the fault in such cases.

## Residual Current Devices (RCD)

RCD is the number of automatic switches found in the main switch. Their main role is to monitor currents in your premises, hence preventing fire outbreaks or shock. Anytime these switches detect imbalanced power flow, they respond by tripping.

For safety, you do not want to install or repair the main switch as a DIY project; always call a professional electrician to work on it.

# Electrical Sockets

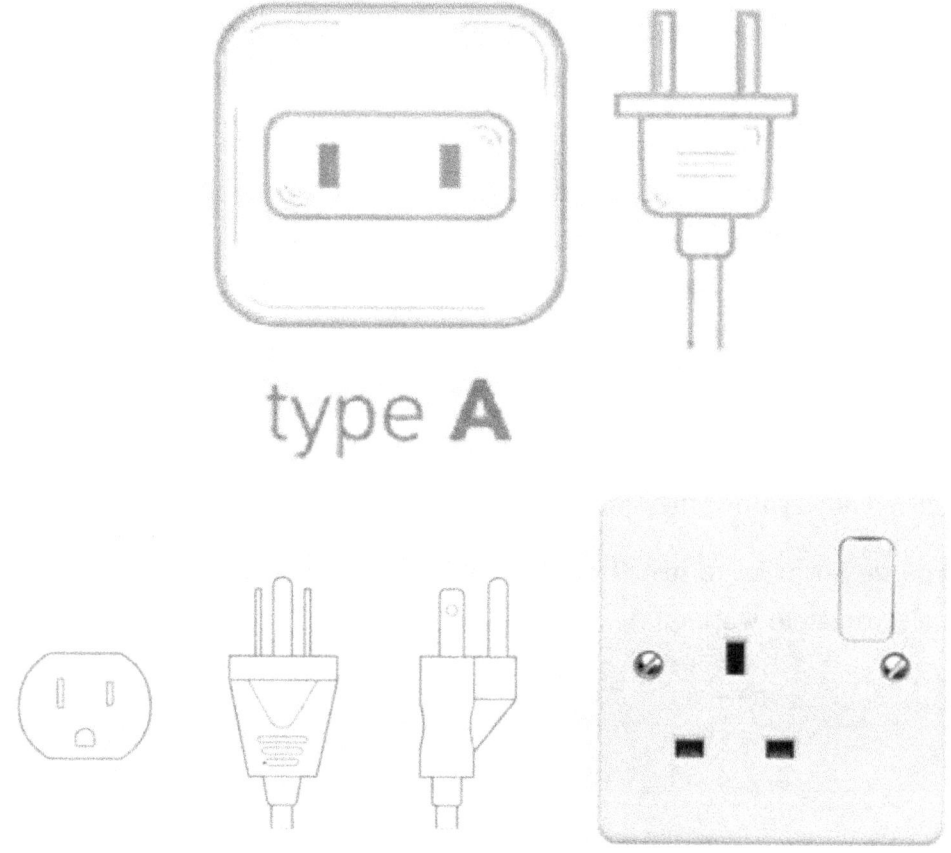

Type B socket                Type G socket

Also known as electrical outlets, power sockets, or wall plugs, electrical sockets are appliances used to plug in electrical devices to give them the power they need to operate. They have at least two slots with hot and neutral wires each. When the prongs of an electrical appliance insert into a socket, the live wire carries current to the appliance while the neutral one returns it, thereby completing the circuit.

Some power sockets have a grounding slot (a third slot) for safety as it diverts power flow in case of short-circuit, while other wall plugs have an additional slot (a fourth slot) with a hot wire for appliances that need more than the standard current.

If you are from the United States, you will have to choose between type A or type B sockets and plugs. Type A sockets have two slots and can only fit type A plugs. This socket type has an outlet of 15A and can offer 100 V to 127 V.

The type B sockets come with three slots with a socket outlet of 15 A; they offer 100 V-127 V.

This type of socket can fit both type A and type B plugs.

The standard frequency is 60 Hz in the USA, and the voltage is 120 V; hence, you don't need a voltage converter as most appliances are labeled "INPUT: 100 V- 240 V, 50 Hz- 60 Hz.

In the UK, you will only find type G electrical sockets. These socket types have three slots, with an outlet of 13 A; they accommodate 220 V- 240 V. Type G sockets can only fit with type G plugs.

Some other terminologies used to refer to the basic electrical tools that you might have to use when wiring your home are:

## Non-Contact Voltage Tester

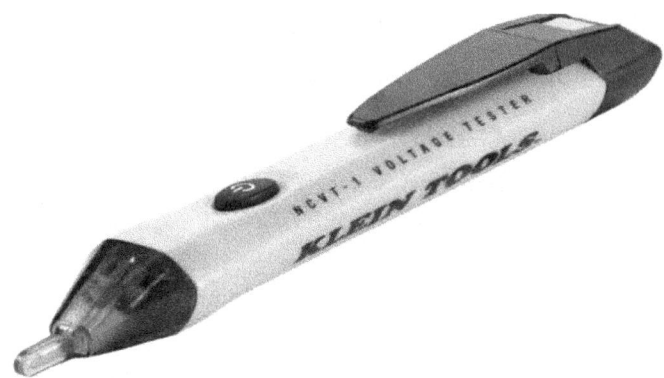

Although there are different types of electrical testers, a non-contact voltage tester, also called an inductance tester, is the easiest to use and safest type there is. It resembles a small wand with a tip at the end and tests voltage in devices and wires.

When you put the tip of an inductance tester on a wire, cable, or any electrical outlet, it will produce a buzzing sound or a red light to indicate the presence of voltage. Although this type of tester doesn't give the exact voltage amount, it is a basic requirement for shock prevention during wiring.

Non-contact voltage testers use a battery; hence, always make sure to charge it before use and test it using a live outlet, such as a switched-on socket, to ensure your tool is in good shape.

## Insulated Screwdriver

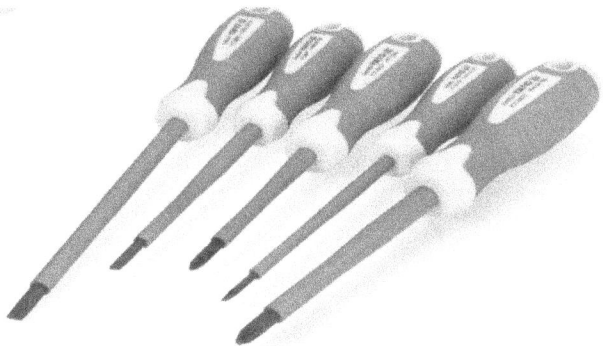

An insulated screwdriver is an electrical tool covered with non-conductive material on its handle and shaft, leaving only the tip bare. This insulation is for your protection from faults caused by live wires and for protecting delicate electrical appliances from damage caused by electrical short. Like any other screwdriver, insulated screwdrivers fasten or unfasten screws by rotating them at the head.

Different types of screws will require different types of screwdrivers, and for you to carry on different projects, you will need to have a set of screwdrivers that have insulation of up to 1000 V.

Also, it is best to buy electricians' screwdrivers with magnetized tips as it will make your screw driving process easier by ensuring the screws do not fall, and their heads easily align with the tips of the screwdrivers.

## Wire Stripper

A wire stripper is used to cut the insulation coating on electrical cables without damaging the copper wires. You will need this tool in cases where you will be repairing/replacing wires or when you will be stripping wires to connect them with terminals or other wires.

Wire strippers are made from steel and are available in different sizes and shapes. Their handles have a rubber coating (to help with grip) and can be curved or straight in shape.

This tool has serrated teeth designed for the standard wire gauge, i.e., 10-20 gauge. Copper wires can be stranded or solid, with the stranded ones slightly larger than the solid ones, which is why wire strippers have two numbering sets showing the teeth (holes) sizes. For example, the hole made for 12-gauge solid wire is the same hole made for 14-gauge stranded wire.

It is essential to match the wire gauge with the right hole because choosing a big hole means it will not remove the whole insulation coat while choosing a small hole might tear the copper wire.

Once you have identified the right tooth size for your wire, open the wire stripper, put your wire in place and gently press the stripper's handles together to their initial position. If this action does not fully cut the insulation, gently twist the wire or rotate the wire stripper, i.e., around 45 degrees to one side and back. After cutting the insulation, pull your tool towards the nearest end to expose the wire.

# Electrical Tape

Also called insulation tape, electrical tape is a product that has high resistance to corrosion, abrasion, and moisture. Its primary use is to insulate electrical wires, preventing the transfer of current to people and other devices, a condition that can lead to shock or fire accidents.

If you need to do a repair or a new wiring project, electrical tapes will help you color-code your wires (since they are available in various colors), making your work easier.

Normally, the back electrical tape is used for installation, while the other colors are used as 'phase tapes" on black insulated wires to indicate their purposes. i.e., red is used as phase B tape to indicate low voltage wires, brown is used as phase A tape to indicate high voltage wires, and green usually indicates earth/ground wire. Yellow is used as phase C tape to indicate high voltage wires (in the USA) or phase B tape to indicate low voltage wire (in the UK).

Although electrical tapes are available in different materials, like rubber, vinyl, varnished cambric, and mastic, vinyl insulation tapes are the best type for home wiring purposes. This is because it has an abrasion-resistant PVC backing, making it flexible and long-lasting. Also, it is the best material for moisture prevention in wires and appliances.

**Electrical Pliers**

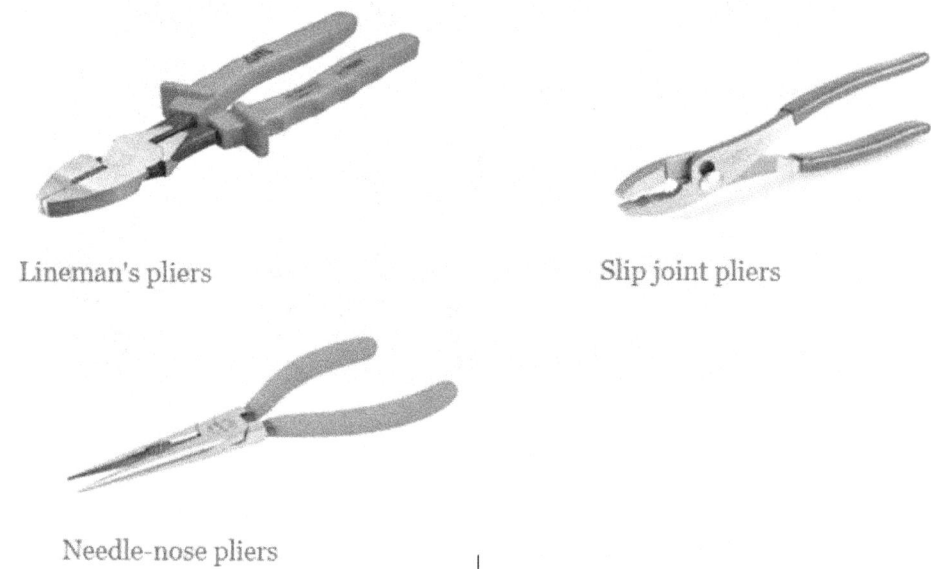

Lineman's pliers

Slip joint pliers

Needle-nose pliers

Electrical pliers are tools you cannot fail to have in your toolbox because they have several applications. There are different types of electrical pliers, but the most basic ones are:

1. Lineman's pliers: These pliers-types have insulated handles that protect users from shock and provide a firm grip when using. You can use it to cut, grip, bend, straighten, pull or push wires, small electrical metals, and cables. You can also use it to strip insulation from wires but not as effectively as wire-strippers.

2. Slip joint pliers: These fasten or loosen bolts and nuts. You can also use them to cut soft nails and wires, loop, and crimp metals.

3. Needle nose pliers: As suggested by the name, these pliers are needle-like at the nose and used for twisting and bending fishing and wires. Its use is common in areas where Lineman's pliers cannot work properly —corners are a good example.

Make sure you have at least the three mentioned electrical pliers for your wiring projects.

## Flashlight

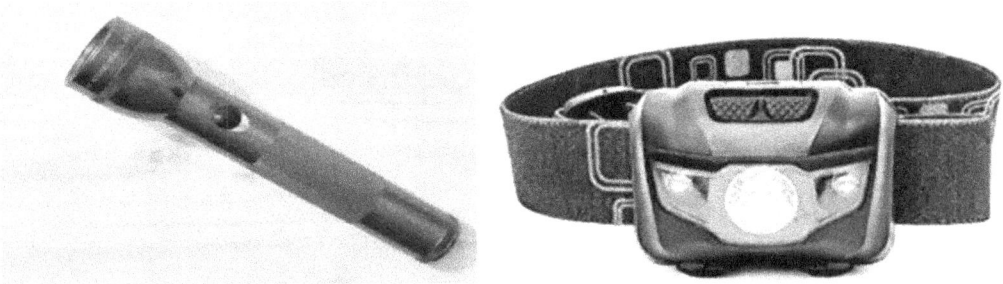

If you ask any electrician, they will tell you that you should never do any wiring project in an area without proper lighting. Some places, such as the basement, ceilings, or walls, might not have the best lighting, and to be on the safe side, always have a well-charged work light, headlamp or flashlight in your toolbox.

## Box Cutter/ Utility Knife

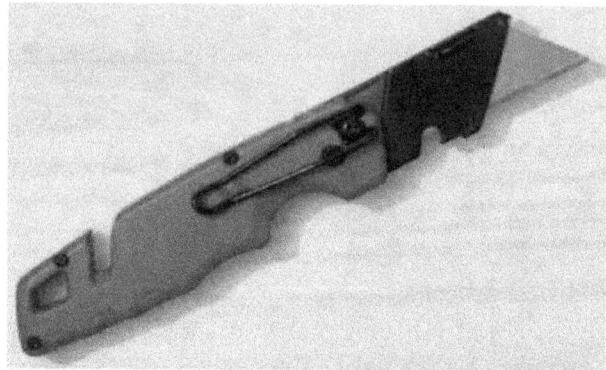

A utility knife might come in handy during wiring. For example, you might need it when cutting electrical tape, sheathing, or opening boxes made of cardboard. As a result, a box cutter is an essential tool you can't miss in your toolbox.

# Fish Tape/Electrician's Snake/Draw Wire

A fish tape is a thin but long steel wire secured in a donut-like wheel. This tape gets its name because its main use is to "fish" wires through conduits, an action that is almost impossible to do with bare hands.

To use it, press the release button near its handle, then unwind the tape as you feed it into a conduit. When the end with the hook reaches the other side of the conduit, let someone help attach the wire to the hook and then pull the fish tape to the starting end.

To return the fish tape into its wheel, hold it (the wheel) with one hand at the center and then turn its handle with your other hand.

When using this tool, always have someone help you pull the tape or attach the wire. As one person pulls the tape wire, the other person should be feeding the wire into the conduit while applying lubricant to avoid damaging the insulation on the wire (a situation that can be very dangerous).

Always untangle wires before fitting them into the conduit to ensure they flow easily. Also, always pull the wire slowly, i.e., pull 3 to 4 feet, stop for a few seconds (those preferred by the person feeding the wire), and then continue pulling. Pulling without consulting your helper can cause injuries to their fingers and nick the cable.

## Electrician Hammer

Not all hammers are ideal for wiring. Electrician hammers have high-density, rubber-coated fiberglass (on the shaft) to absorb shock and narrow faces and long necks that enable them to reach tight places, like outlet boxes. In addition, the materials used to make their handles help give you a firm grip.

You will need an electrician hammer to secure nails and wire staples into place when installing electrical devices. You will also use this tool to remove nails and wire staples when removing or replacing electrical appliances.

However, never use an electrician hammer on wires carrying current as it can be dangerous. Also, avoid using it on devices that need nailing down with a lot of force because it can damage the handle or shaft.

# Tape Measure

You will be doing a lot of measuring, e.g., determining the length of conduit needed or the location of switches, sockets, and bulbs. This means you need to have a tape measure in your toolbox.

Look for a tape measure with a magnetic tip; this will allow you to hook it up on a magnetic material, making your work easier. Also, make sure your tape measure has a heavy-duty blade so that you can extend it as far as you want without cases of breaking or collapsing.

## Wire Connector

Wire connectors are devices used to make stable and secure wire connections during wiring projects. For example, if you have a short wire issue, you will need a wire connector to connect the existing wire to an additional wire, thus having the required length.

Some people use electrical tape in place of wire connectors. Never take this risk because wire connectors are safer than electrical tapes. Their unique design helps prevent any contact between the connected wires and other wires in the circuit or any other object that can lead to electrical short.

Each wire connector has a minimum and a maximum number of wires it can hold, a characteristic determined by wire gauge, e.g., a wire connector might have "2#14 wire" and "4#12 wires" indication on it. This means it can hold a minimum of 2-14 gauge wires and a maximum of 4-12 gauge wires.

Different types of wire connectors come in different sizes and shapes, but the basic ones are:

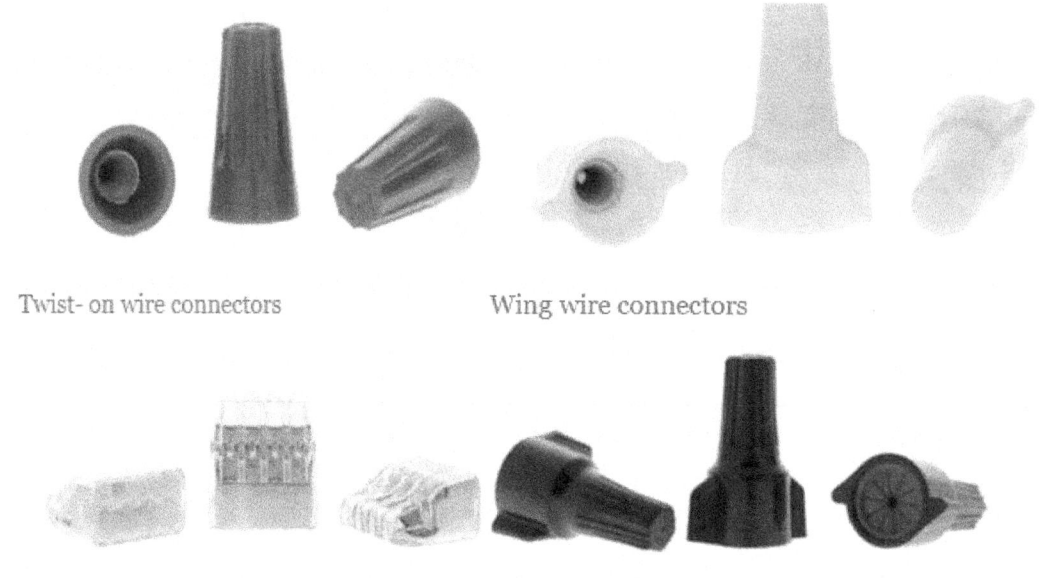

Twist- on wire connectors   Wing wire connectors

Push- in wire connectors   Waterproof and weatherproof wire connectors

- Twist-on wire connector: These are the type of wire connectors mostly used to connect wires at homes because it is simple, economically friendly, and reusable (if not damaged). It has ridges on its cap barrel that allow a firm grip when twisting it onto wires. It also has a square-cut spring on the inside, whose function is to grip and hold wires together after they are connected. It is used when installing/replacing fixtures, switches, and receptacles.

- Wing wire connector: It has a similar application as a twist-on connector, but different in that its tabs/wings on the cap barrel allow for easier and quicker installation/removal, making it suitable for projects that call for many wire connections.

- Push-in wire connector: These connectors are a bit expensive compared to wing wire and twist-on wire connectors. However, it is a quick and easy-to-use tool as all you have to do is insert wires into it without needing to twist it or the wires. Some push-in wire connectors have translucent housings, which allows you to confirm proper wire insertion. In contrast, others have check ports, which allow you to test if the connections are properly functioning without removing them.

- Waterproof and weatherproof wire connectors: They resemble the wing wire connectors, but they have flanges/fins on the cap openings and silicone covers on the caps to protect the wires from moisture, dust, and corrosion. They are suitable for outdoor sound systems, light fixtures, and outlets connections.

Having known the basic wiring terminologies and tools you might require, let's look at some of the safety precautions you need to know and keep in mind during all your wiring projects.

# Chapter 2- Wiring Safety Precautions and Tips

Your safety is not something you can compromise. You should always take precautions when doing any wiring project. After all, whenever you're working on any electrical appliance, you put yourself, the people in your homestead, or your property at risk.

However, you can minimize the chances of accidents by taking proactive actions when handling any electrical devices, equipment, or appliances. It doesn't matter how simple a project seems; never forget that electricity can be dangerous, and one simple mistake can cost you a lot, if not your life.

Some safety tips you should always keep in mind are:

## Cut Off Power Supply

Before starting any wiring project, turn off the main switch and circuit breaker at the consumer unit to cut off the power supply. This is because you should never work on any electrical appliance while the current is flowing. Always use your voltage tester to confirm that electrical connections and wires are dead.

Also, let everyone in your home know that you are doing a wiring project, and the power is off for a reason. If possible, leave a note on the consumer unit saying no one should turn the power back on.

**Be Cautious of What You Touch**

Avoid touching anything you don't know or unmentioned in your project guide. More importantly, never touch your gas or plumbing pipes (it doesn't matter if you are wiring or not) because they are mostly used to ground current.

## It Is Better To Have a Long Wire than a Short One

Anytime you are cutting a wire, make sure to cut a few inches longer than the length you need. It is better to work with a longer wire because you can cut it to the appropriate length after installation instead of having a short one. Having a short wire can lead to poor connection.

If you mistakenly cut a short wire, you can fix it by adding an extension of at least a 6-inch wire.

## Never Use a Steel or an Aluminum Ladder

If you need to use a ladder for your project, do not use steel or aluminum ladders. These materials are good conductors of electricity, and they will use your body as the ground path, causing electrocution.

Instead, use ladders made of wooden, fiberglass, or bamboo, as these materials are non-conductors.

## Avoid Water

Water is a good conductor of electricity, and if your hands are wet or if you stand on water and touch any device or wire with a live current in it, the water will transmit this power through your body to the ground, causing electrical shock.

The best way to avoid this is to ensure all your electrical appliances are away from water. Also, always make sure your hands and the ground you are standing on are dry before working on any electrical project.

If you have any electrical devices in water-prone areas, perhaps the hairdryer in your bathroom, unplug them when not in use.

For outdoor water sprinklers, always keep the power lines as far away as possible. If water encounters these lines, electrical current will flow through the water, and you will get electrocuted when you step on it (the water) or touch it.

This also explains why you should never use water to put off an electrical fire; instead, always use a fire extinguisher in case of fire accidents.

## Always Use the Right Wire

Wrong wiring is a common electrical mistake you should always avoid. Wrong wiring can be a small thing like leaving wires unprotected or a big thing like using them for the wrong purpose, e.g., using the hot wire as the ground wire. These mistakes can put your life and the life of those around you at risk; hence, avoid them through the following ways.

## Using the right wire

There are different wire color codes put in place. These codes vary from one country to another; hence, it is important to keep in mind the code used in your country.

In the States, the direct current (DC) power wire code is red for positive (hot) wire, black for negative wire, and grey or white for the ground wire.

The alternating current (AC) power wire code has two categories, i.e., one phase (single-phase) and three-phase. The Single phase has 3 or 4 wires, i.e., 2 hot wires, one neutral wire, and one ground wire. Each live wire (phase wire) carries 120 volts; hence, the 2 wires allow 24o volts of electricity to flow. This phasing type is mostly used in homes since only typical loads, e.g., heating or lighting, are used. The three-phase wiring is used in industrial settings to operate machinery.

In the United States, the wiring color code for one-phase wires is Black and red for the 2 hot wires (sometimes blue is used), white or grey for neutral wire and bare wire, green with yellow stripe or green for the ground wire.

However, the USA National Electrical Code only mandates the ground and neutral wire codes. This means any other wire colors can be used for the hot wire.

The UK uses the International Electro-technical Commission (IEC) wiring code for AC circuits. This wiring color-code rule was introduced as IEC 60446 and later merged to IEC 60445 (in 2010).

The old wiring color code was red for hot wire, black for neutral wire and, green/green with yellow stripes (pre-1977 IEE/ pre- 2044 IEE, respectively) for earth wire. The new IEC wiring color code is brown for live wire, blue for neutral, and green with yellow stripes for the ground wire.

If you bought or inherited your home, it is important to know the color code used to avoid using the wrong wire when repairing or installing new connections. If your home uses the old wiring color code, you can have a professional replace it with the new one or label the wires before working on them to avoid confusion.

The wiring color code only applies to current-carrying wires usually sealed together with grey or black cables to protect people from electrical accidents. However, it is still important to label any wires that people might interact with to inform them of the amount of power present and warn/remind them of the possibility of accidents.

Always treat all wires as live wires because they all have electricity in them, i.e., the live wire carries power to the load, the neutral wire carries current back to the circuit breaker to avoid overpowering the load. The earth wire provides a path for the stray electricity.

When it comes to data/network cables, people often assume they are harmless, which is not always true. Some of these cables have electricity passing through them that can cause accidents.

For example, you have to connect your phone to a cable to get its power and, if this cable is frayed or cut, it becomes a fire or electrical shock hazard. However, data wires are colored based on the manufacturer's standards or needs. Hence, you should never assume they are made using electrical standards.

## Ground the 3rd slot

If a wiring project calls for an outlet with three slots, it is better to avoid problems by connecting the earth wire. Do not use a 2 slot outlet in place of 3 slot outlet because, as stated early, the earth wire offers the stray current a path to the ground, allowing surge protectors to guard appliances against damage and reducing chances of electrocution and electrical shocks.

**Always put on electrical personal protective equipment (PPE)**

Another important electrical safety tip you should always keep in mind is never to do any electrical project without electrical PPE. The gear helps protect you from potential electrical hazards.

The common PPE you should always have on you when doing electrical projects are:

**Rubber insulated (voltage) gloves**

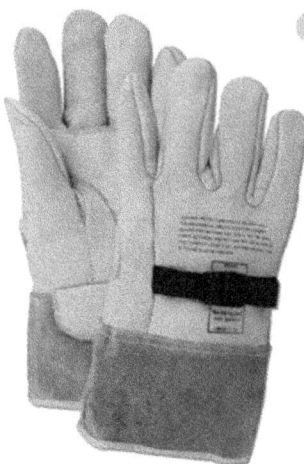

These gloves protect you from electrical shock, which occurs when you accidentally drop tools on electrical equipment or when anybody comes into contact with electrical wires. They also help protect against arc flashes, i.e., an electrical hazard that happens when current travels through air gaps, releasing intense energy that can explode.

These gloves are made of leather covers that protect from abrasion, puncture, or cut damage on the voltage-rated layer (actions that can reduce the gloves' effectiveness). To reduce discomfort while wearing rubber voltage gloves, buy FR liner gloves to wear before wearing the insulated ones.

Rubber insulated gloves are electricity tested and rated from class 00 to class 4, depending on the voltage protection they provide. This means you should choose gloves that offer protection from 240 V (the common amount of voltage in standard home wiring and appliances).

**Footwear**

Electrical hazard (EH) shoes or boots are made from non-conductive materials that keep people from grounding stray electrical current. They can prevent the flow of up to 600-volt current in a dry area.

These shoes/boots also protect sharp, rolling, or heavy objects and prevent falls on uneven services. However, it is unsafe to put on wet EH boots or shoes with worn-out soles or a metallic material in or under them.

These boots or shoes have a tag or an "EH" imprint to differentiate them from other types.

## Dielectric safety goggles/glasses

They are safety glasses made with a non-conductive frame. They do not have any metallic parts, making them ideal for protecting the eyes from injuries like arc flashes or sparks, dust, and flying objects during wiring.

Common safety goggles you can choose from are Pyramex Cappture, Pyramex Trulock, Crews Law 2, Pyramex Endeavour Plus, and Bolle Prism.

## Arc flash suit/ jacket

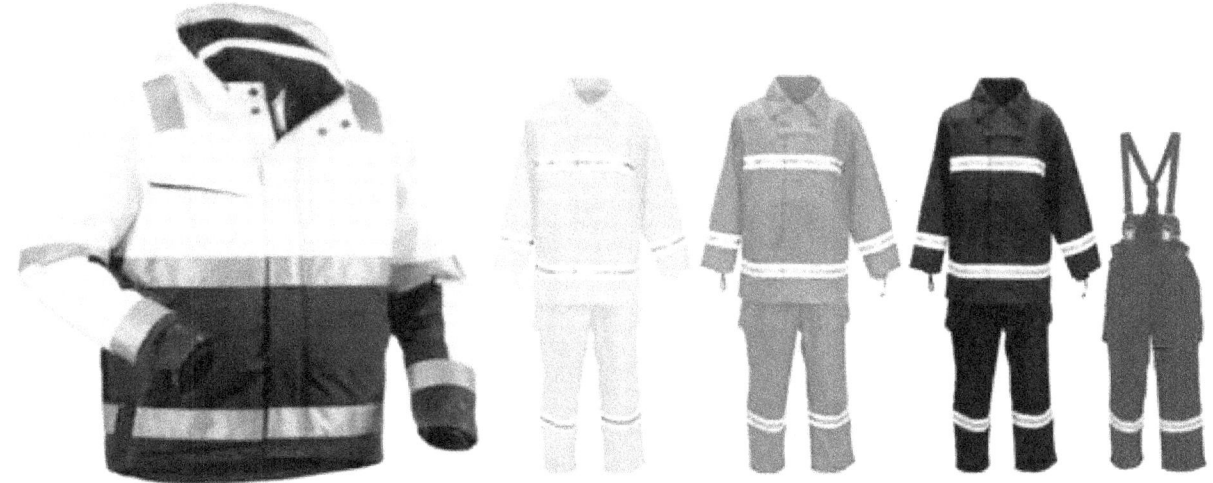

Always put on an arc flash jacket or suit anytime you are working on electricity to ensure your safety. The purpose of arc flash suits or jackets is to protect against electrical shock and arc flashes.

You will only need a hearing protector during noisy situations, like when you're working on your wiring project while construction is underway. Also, put on a safety helmet during outdoor projects to protect yourself from falling or head height objects.

It is better to be safe than to lose your property, to be injured or dead. Therefore, always have the above safety tips and precautions at the top of your mind before and during any wiring project.

Let's now look at some simple wiring projects you can do in and around your home.

# Chapter 3: Simple Wiring Projects for Beginners

Yes, you have basic wiring knowledge, but this does not make you an expert in wiring. Never try some wiring projects since they require training and experience. For example, do not install a consumer unit installation/repair or gas line repair.

That said, some wiring projects are easier to do, and with the knowledge and safety tips you have learned, you can try them out. Some of these projects are:

## 1: Replacing a One-Way Switch

Before you replace your switch, you have to examine it visually to ensure it is a Single-pole switch. After removing the switch cover plate to see the switch, you will see two screw terminals (at the sides of the switch), both connected to hot wires, i.e., the upstream one bringing power to the switch and the downstream one taking power to the appliance.

However, this might not always be the case as this appearance varies depending on the initial wiring first done in your home and the circuit configuration used, but never forget that both wires are live. Remember that the hot wires in a one-way switch are interchangeable. Hence, it doesn't matter to which screw terminal they connect.

In addition, the new one-way switches have a third screw terminal (green in color) for grounding the circuit's system. If your home has the old one-way switch types, it might not have the 3rd screw terminal but, it is important to include this connection in your project by pig-tailing it (the screw terminal) to the grounding wire.

Pigtail is the process of combining many wires, making a single conductor that you can connect to a device, e.g., a socket or a switch. You will also use this technique to connect ground wires to metal electrical boxes or lengthen short wires.

For this project, you will need a screwdriver, wire stripper, non-contact voltage wire, pliers, a one-way switch, wire connector, and a grounding pigtail wire (if needed).

### Step 1: Turn off power

Switch off the circuit breaker to cut off the power supply to the circuit you will be working on — always do this at the start of each project. If your circuit system uses fuses (instead of breakers), remove the appropriate fuse (the one for the circuit you will be working on) by unscrewing it.

However, do not trust the circuit labeling on your consumer unit until you test to confirm that power is off (which is the second step).

## Step 2: Test to confirm that power is off

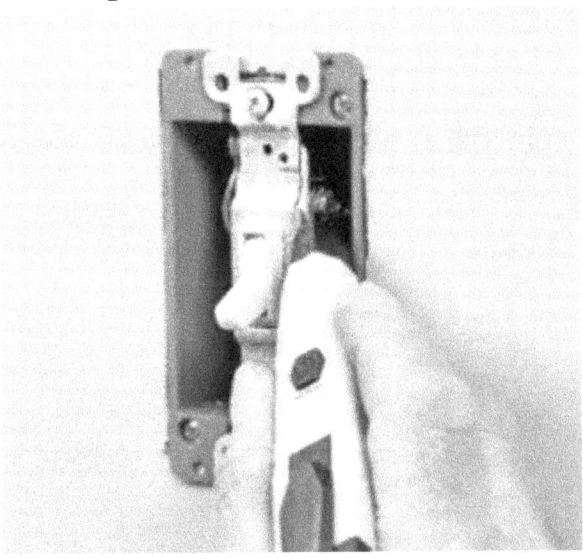

Using your screwdrivers, unscrew the screws on the existing switch plate. Slowly and carefully remove this plate from the wall. Using your non-contact voltage tester, test to confirm that electrical power is off by placing its tip on all the wires in the electrical box. Also, use your tester to test for power in the screw terminals.

If your non-contact voltage tester produces sparks or some light on the handle, current is still flowing, and you should go back to your consumer unit and turn off the right circuit breaker. Again, test to confirm that the power is off.

## Step 3: Remove the existing switch

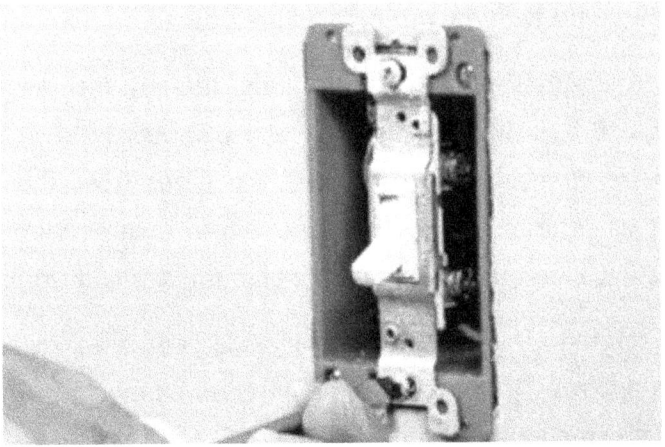

Use your screwdriver to loosen the screws holding the existing switch in place and carefully pull it. This action will leave the switch hanging, only suspended by the wires connected to its screw terminals.

Note that if the existing single-pole switch is the dimmable type, its appearance will differ from the standard type. For example, it is larger, and its connection is made using a wire connector or leads (instead of screw terminals). In such cases, make sure you have the right replacement switch.

## Step 4: Disconnect the electrical wires from the screw terminals

First, confirm that the current is off before doing anything and then inspect to confirm that each screw terminal has one wire under it. You will probably see a black wire on one terminal, a red or white wire on another, and a green wire on the third terminal if a grounding system exists.

If you notice a white wire used in one of the terminals, it probably means that it is a loop configuration, i.e., the switch is the last connection appliance in that circuit. If this is the case, the white wire is a live wire and should have black or red tape. If not labeled, make sure to do it before installing the new switch.

If the switch box has other wires in it, you should not touch them as they are just passing through the box and have nothing to do with the project you are doing.

Use your screwdriver to unscrew the screw terminals, thereby disconnecting the wires. If the wires connect to the switch by a wire connector, push your screwdriver or a small nail into the "release slot" on the wire connector to disconnect the wires from the connector. Electricians regard wire connectors as inferior as they are less secure than screw terminals. Hence, when installing the new switch, it's best to connect its wires to the screw terminals.

If you are replacing a single-pole smart switch, i.e., one controlled by wireless devices, e.g., a phone, you will need a neutral wire to uphold the wireless connection. In such scenarios, install these connections as instructed by the manufacturer.

## Step 5: Installing the new switch

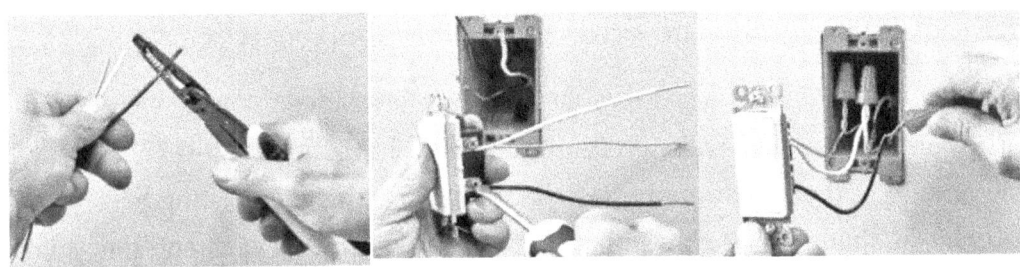

Start by connecting the ground screw terminal to the ground wire. If one ground wire is present, directly attach it to the screw terminal. If two ground wires are present, you will have to connect them by putting a pigtail in the ground screw terminal and attaching the ground wires to the pigtail using a wire connector.

*How to make a pigtail connection*

A pigtail connection is a necessary step for this and other projects. Hence, it is important to learn how to make it through the following steps:

- Use a wire cutter to cut 6 to 8 inches of a scrap wire of the same wire gauge and color as the wires you want to connect (for your ongoing project, it should be green or green with yellow stripes).

- Use a wire stripper to remove an insulation bit of about 0.75 inches from the end of each circuit wire being pigtailed. Some devices have strip gauges on their sides, indicating the length of wire insulation to be stripped. Ensure you check and follow those measurements, if any.

- Using your needle-nose pliers, connect one end of the pigtail to the screw terminal (this should be the grounding terminal) by looping it to the right. Use a screwdriver to tighten the screw terminal and ensure the wire loop is secure and the wire is only on the screw shaft with no bare copper exposed.

- Connect the exposed circuit wires to the other pigtail end with a wire connector. Pull the wires, i.e., the pigtail wire and the circuit wires to ensure that the connection is tight and secure. Trim off any excess bare wire to ensure you do not leave any copper wire revealed.

- If you are using a grounding pigtail connection with a metal electrical box, loop one end of the pigtail to the green screw and thread it into the screw opening in the electrical box (at the back). Attach the other pigtail end to the circuit grounding wires. Some metal electrical boxes will require two pigtails. In such cases, you will have to attach one pigtail to the appliance, the other pigtail to the electrical box, and then connect the two free ends of the pigtails to the circuit wires with a wire connector.

After connecting the grounding wires to your new switch, you can now move on to the two hot wires. Start by inspecting their ends to ensure they both have about 0.5-0.75 inches of exposed wire. If one or both of these ends are in bad conditions, trim off the existing bare wire, then strip an insulation bit of about 0.75 inches to have new bare wire. Bend each end into a hook-shaped loop and wrap them on the screw terminals (each wire on one terminal) in a clockwise motion. Tighten the screws and tug on all wires to ensure all connections are secure.

## Step 6: Finish up the project

Gently tuck all the wires into the switch box and secure the switch on the box by mounting it with its screws (normally 2). Also, mount the switch cover plate into place.

Turn on the circuit power by switching on the circuit breaker or securing the fuse back into its place. Test if your new switch is operating properly by turning it "ON" and "OFF."

## 2: Wiring an Electrical Socket Outlet

Before starting this project, you need to understand that sockets outlets have two types of wiring. The first one is when the outlet is in the middle of a circuit. In this case, the electrical box for this outlet will have 2 or 3 wires entering it, i.e., one wire that brings power into the box, and the other (1 or 2 wires) that carry power to other fixtures or outlets in the circuit.

Here, the wire bringing current to the box can connect to the live and neutral screw terminals. On the other hand, the ongoing wires can be connected to the other screw terminals on the receptacle, allowing all power to pass through the outlet and flow to the other fixtures/outlets in the circuit.

Another way of wiring socket outlets (which is better and most followed) that fall in the middle of a circuit is by using a pigtail. One end of the pigtail connects to the live and neutral screw terminals while the other end attaches to the circuit wires, i.e., the one bringing current to the outlet and the one/s carrying current to other outlets or fixtures within the circuit.

The other method of wiring socket outlets is when they come at the end of the circuit. This method is easy because only one incoming wire in the socket box needs securing to the live and neutral screw terminals.

The tools you need for this wiring project are a wire ripper, screwdrivers, needle-nose pliers, wire connectors, a utility knife, scrap NM wires, and a socket outlet with a cover plate.

### Step 1: Turn off power

Turn off the power by switching off the circuit breaker or unscrew the fuse (if the wiring system uses fuse panels) that controls the power to the circuit on which you will be working. If you are unsure of the circuit, completely switch off the power to your premises in the main switch.

### Step 2: Test the switch outlet for voltage

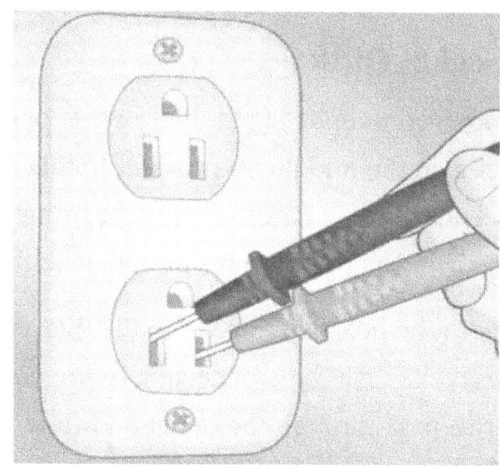

During wiring projects, never assume anything, and even if you have switched off the power, you still need to test the voltage. Insert your non-contact voltage tester into the socket holes to ensure the wires are dead. If your tester blinks or produces sparks, it means the current is still flowing, and you cannot continue with your project until you are sure there is no voltage in the wires.

If you are from the UK, you will have to consult an expert to confirm if your house has ring circuit wiring. If that is the case, you will have to hire a professionally certified electrician to do this project because ring circuit wiring is complicated. You do not have to worry about the ring circuit wiring system if you live in the US.

## Step 3: Remove the existing socket

Use a screwdriver to remove the screws on the outlet cover plate. After that, remove the socket outlet from the socket box by unscrewing its screws, then loosen the screws on the terminals to detach the socket from the wires.

## Step 4: Identify the terminals on the new socket

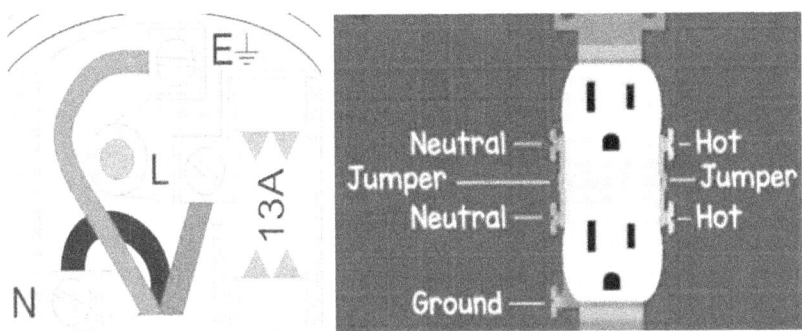

UK socket outlet terminals     USA socket outlet terminals

Most household socket outlets have three terminals for connecting the hot, neutral, and ground wires. In the USA, the brass terminal is for the hot/live wire, the silver terminal is for the neutral wire, and the green terminal is for the ground wire. In the UK, "L" indicates hot wire, "N" indicates neutral wire, and "E" indicates earth/ground wire.

If your socket has more than three terminals, it might be because of these reasons:

- In the UK, you might have to connect two wires of the same wire type, i.e., hot and neutral wires to their appropriate terminals (each wire on one terminal). However, you do not have to use all the terminals in such cases because you can connect two wires of the same type to one right terminal using wire connectors.

- In the USA, a socket with 2 outlets has 2 metal tabs for connecting the 2 live and 2 neutral terminals (each wire type connected to one metal tab using a wire connector). If your wiring system offers only one set of wires (a hot, neutral, and ground), you can connect it to the appropriate terminals in either socket and, this connection will power the 2 sockets.

## Step 5: Prepare the wires

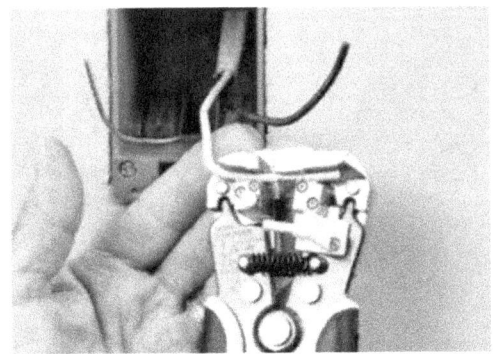

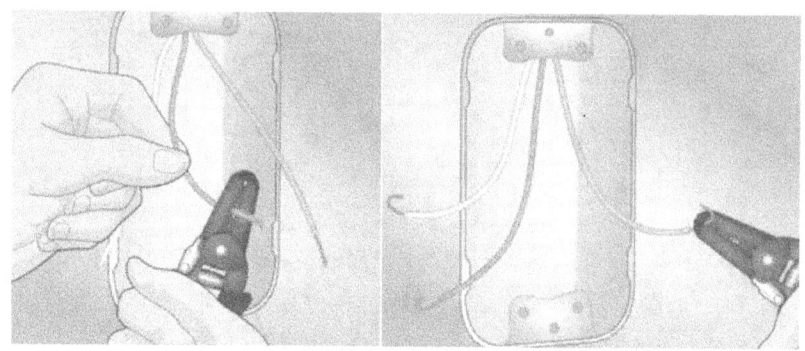

Wires emerge from the wall into the electrical box, through its sides or back, and are secured in place by pressure-fit braces or metal wire braces. Only a small part of the outer covering of the NM cable should be visible in the outlet box, and only 6-7 inches of the conducting wires should extend into the socket box. If these wires are longer than the stated length, you should trim them with a wire cutter because long wires can be unwieldy and challenging to fit into the outlet box.

If the wires are nicked or frayed, trim off the damaged part, then use your wire stripper to strip off an insulation part of about 0.75 inches (2cm). However, most socket outlets and wire connectors have imprinted instructions on how many inches to strip off. Always check the back of your socket or wire connectors for the manufacturer's instructions and follow them. This step is important because it will help you avoid having exposed bare wires on the screw terminals or wire connectors.

Use needle-nose pliers or wire strippers to bend the bare wire ends into a hook-like shape. Doing this will make your work easier when connecting them to the screw terminals or wire connectors. However, if your project calls for a pigtail connection (e.g., in cases of 2 live wires), you do not need to bend the wires used in this connection as you will only need to push or twist them into the wire connector.

## Step 6: Create a pigtail connection

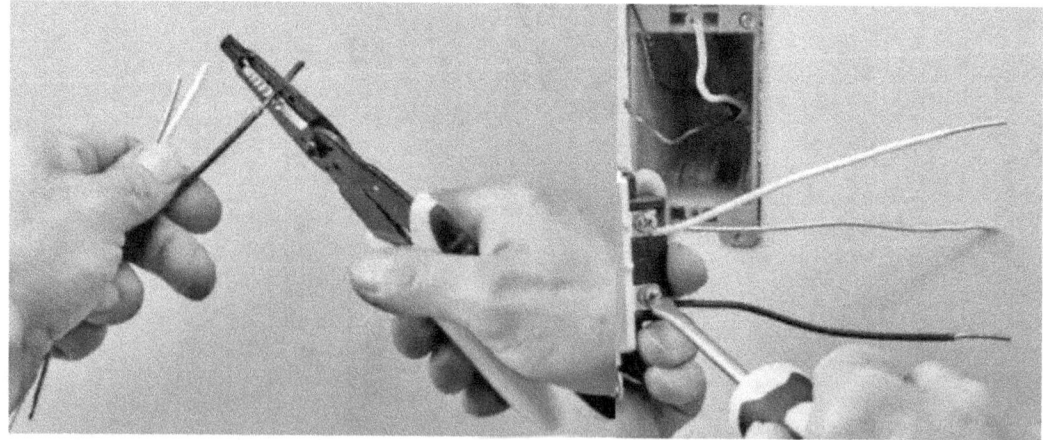

The way you connect the wires to the outlet depends on the number of cables in the electrical box. If there are 2 or more wires of the same type, you will have to connect them to their screw terminal using a pigtail connection. Cut 6 inches of the NM scrap wire of the same type as the wires you will be connecting, i.e., if you have 2 hot wires, red and black (in the USA) or brown and black (in the UK), you can use a scrap wire of any of the 2 wire colors (depending on where you are).

Strip 0.5 inches and 0.75 inches on each end of the pigtail, then attach the 0.5-inch end to a wire connector and bend the other end (using needle-nose pliers) into a hook-like shape as you did for the other wires in the above step. If there is more than one ground wire or neutral wire, you will have to create a pigtail connection for each as you did for the hot wire.

If the socket box has one cable with 3 wires (hot, neutral, and ground), there is no need for a pigtail connection as you will connect the wires directly to the appropriate terminals.

If the socket box is metal, you will need 2 ground pigtails. Connect one of them to the ground terminal on the socket and the other to the ground screw in the socket box, then attach them to the ground wire/s using wire connectors.

Remember to use wire connectors of the same gauge as the circuit wires.

## Step 7: Attach the wires/pigtail(s) to the screw terminals on the socket

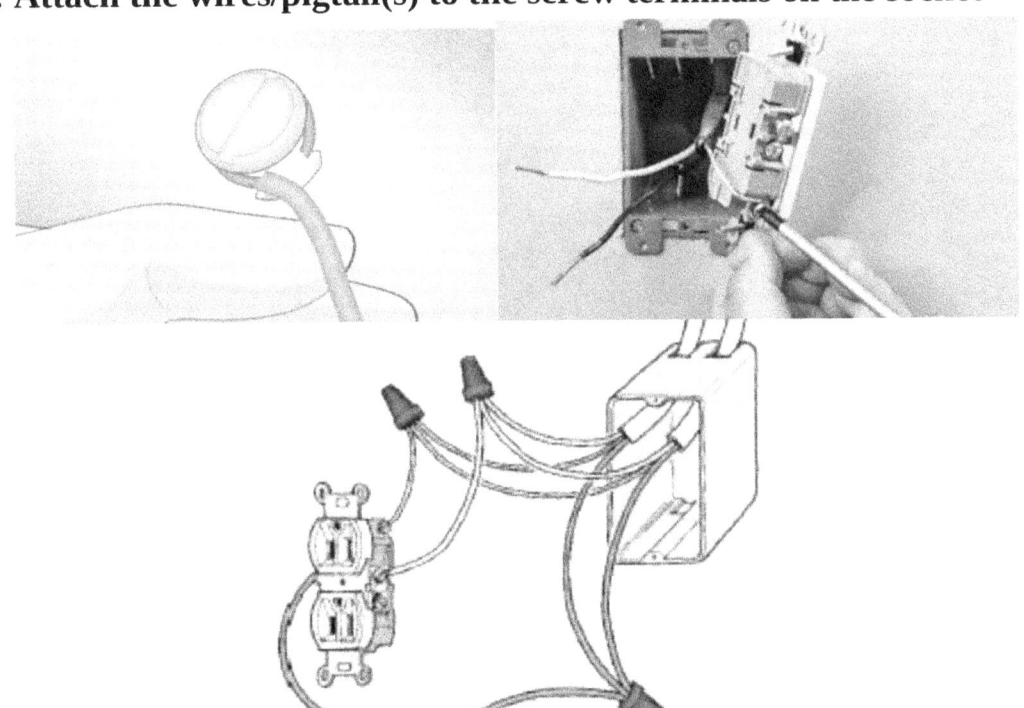

Start by connecting the ground wire or pigtail. Take the hook-shaped ground wire and wrap it around the socket's ground/earth screw terminal (green in the USA and labeled "E" in the UK) in a clockwise motion. This will help close the wire into place as you drive the terminal's screw into the socket. Ensure the bare wire fits snugly on the screw's shank by squeezing it (the wire) using needle-nose pliers. Use a screwdriver to fasten the ground screw.

Attach the neutral wire/pigtail, i.e., white or grey (for the USA) or blue (for the UK), to the neutral screw terminal (silver in the USA and labeled "N" in the UK) in a clockwise motion. Let the wire insulation just touch the neutral terminal to ensure you don't have bare wire left exposed, then fasten the neutral screw into place.

Attach the live wire/pigtail, i.e., black or red (for the USA) or brown/black (for the UK) to the live/hot screw terminal (brass in the USA and labeled "L" in the UK) in a clockwise motion. After making sure you have no bare wire exposed, tighten the live screw.

**Step 8: Tuck the wires into the box or under their plastic retainer**

Some socket outlets have plastic retainers for tucking the wires, but if the one you are working on doesn't have them, you will have to tuck them gently into the socket box by bending them.

After tucking the wires, double-check to:

- Ensure no insulation is in contact with the screw terminals (if any, strip it off) and no bare wire end is exposed (if any, snip it off using a wire snipper).
- Confirm that the wires connect to the right terminals.
- Ensure the wires are secure and tight by tugging on each of them. If you have any loose wires, retighten or reconnect them.

## Step 9: Attach the socket to the socket box

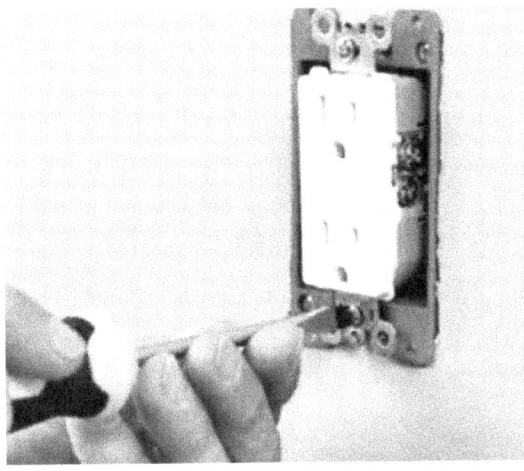

Now mount the socket's top and bottom metal straps on the socket box by fastening their screws. Electricians consider it safer when the ground slot (the D-shaped hole on its own line) is on the top. This is because the ground prong will ground current, preventing a short circuit if an object falls on a plug partially pulled out of the socket (with its prongs exposed). Due to this technicality, attach the socket to the socket box with the ground slot on the top.

## Step 10: Attach the socket cover plate

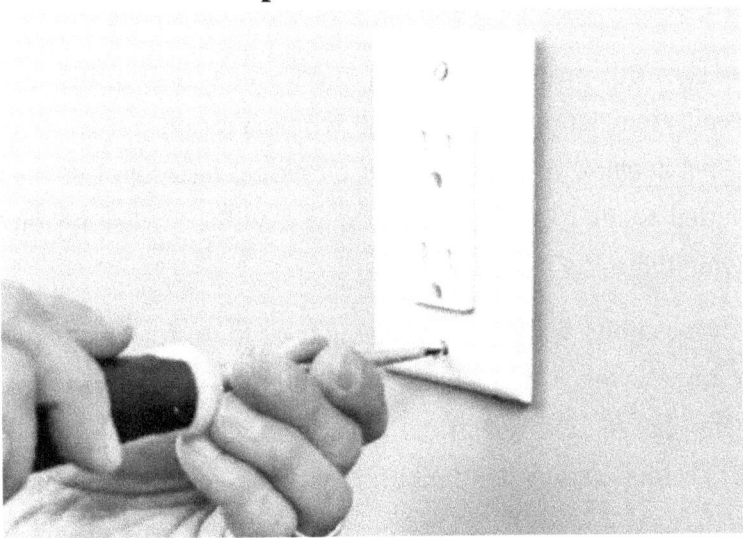

Mount the socket cover over the socket and use its screws to secure it into place.

Now turn on the power by switching on the circuit breaker, reattaching the fuse, or switching on the main switch. Use a tester to check if there is power by putting its tip in the socket holes and then plug in an electrical appliance to the socket outlet to confirm if it is working properly.

## 3: Installing an Outdoor Lamppost

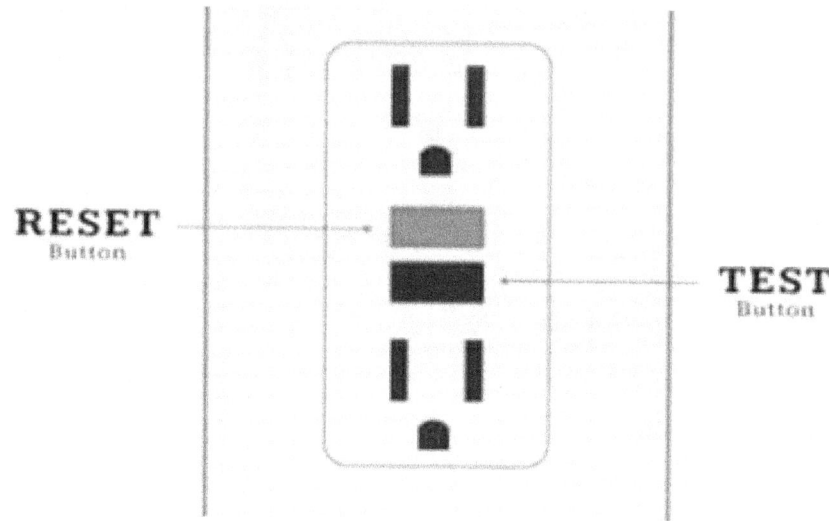

A GFCI outlet

Installing a lamppost at the back or front of your yard is an easy-to-do wiring project that will change how your compound looks and make night movements easier and safer.

However, before you start this project, you need to contact a local utility company to inspect the area you want to install your lamppost for buried wires. You do not want to dig in an area with existing electrical wires because it might lead to electrical hazards. Let the utility service mark all the areas with buried wires near the yard area you will be working on with spray paint or strings so you can avoid them.

When it comes to buying cables for this project, the type to go for will depend on if you want to include a conduit or not. However, it is best to use conduits, as they will provide extra protection to the wires, thereby preventing electrical accidents. To be on the safe side, buy UF (Underground Feeder) cables as they are the best type of cables for underground projects.

Use your tape measure to determine the approximate length of cable and conduit you will need for your project by measuring the distance between your GFCI outlet and the position you want to install your lamppost. A GFCI (Ground- Fault Circuit Interrupter) is an outlet used to detect electrical current imbalance, e.g., when water comes into contact with electricity and automatically shuts power off to prevent damages and accidents.

Look around your home to locate the outside GFCI outlet to your home. It is an easy to locate outlet because of its "Test" (back) and "Reset" (red) button found on its face. If you cannot find it, have a licensed electrician install it, as you should never do any outdoor project without this conducting connection.

Also, have a professional install a power pack on the wall and connect it to the GFCI outlet. A lighting power pack is a device installed in a junction box. Power will flow from the GFCI outlet to the power pack from where you will connect your UF cable. A junction box has a switch for turning the power "ON" and "OFF."

A power pack is a low voltage sensor system that transforms 120-277 VAC (50-60 Hz) into VDC current for occupancy sensors. It has an automatic system that turns loads "ON" and "OFF." This means if you have a given occupancy sensor, e.g., light or motion sensors in the power pack, it will detect the presence of motion or light and respond by switching the load on (in case of motion) and off (in case of light).

With this device, you won't have to switch your lamppost "ON" every night and "OFF" every morning; the power pack will do it for you. However, if you prefer the manual "ON" and "OFF" system, you will have to use the switch in the junction box. A power pack is ideal as it comes in handy when you cannot manually turn the lamppost "ON" and "OFF"—for example, when you are away for a vacation.

In addition to the electrical tools in your toolbox, you will need a shovel, a wheelbarrow, sledgehammer, masonry hoe, trenching shovel, hacksaw, concrete float, lamppost with a lantern, and posthole digger for this project.

### Step 1: Turn off the power and use a tester to confirm if there is voltage

As stated early, you should never undertake any wiring project with the power on. Switch off the circuit breaker powering the outlet you will be working on. If you are not 100% which one it is, turn off the main switch at the consumer unit. Use a voltage tester to confirm no voltage in the GFCI outlet.

## Step 2: Mark your working area

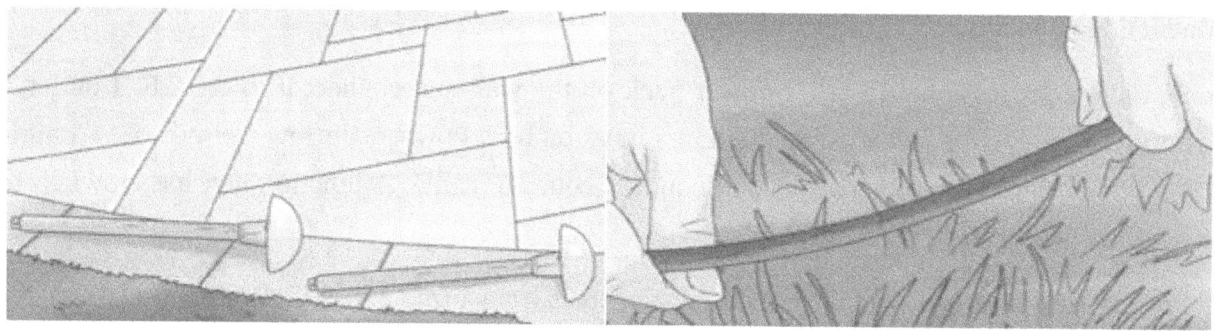

Use a mini flag or a stone to mark where you intend to install the light pole. After that, lay your cable from the power pack to the stone or mini flag. Use string/spray paint (of a different color from the one used by the local utility company) to mark the area where the cable will go through.

Also, make sure to leave at least 4 feet between your working area and any obstacles, e.g., sidewalks, shrubs, or trees, and any mark made by the utility company. If your cable has to cross a footpath, make the string or spray paint markings across the footpath and continue to the pole location.

This will help direct you to the finishing point while avoiding possible accidents caused by obstacles or existing buried wires.

## Step 3: Dig the posthole

Use a round shovel and a posthole digger to dig a 12 inches wide and 22 inches deep hole at the location you marked with a mini flag or a stone. Add 6 inches of gravel to the bottom of the hole and use a trenching spade to make the hole about 18 inches deep, with the gravels at the inner sides of the hole. This important step helps form a firm foundation for your lamppost since soil shifts during dry, rainy, or frosty seasons, making the ground unreliable footing.

## Step 4: Dig the trench

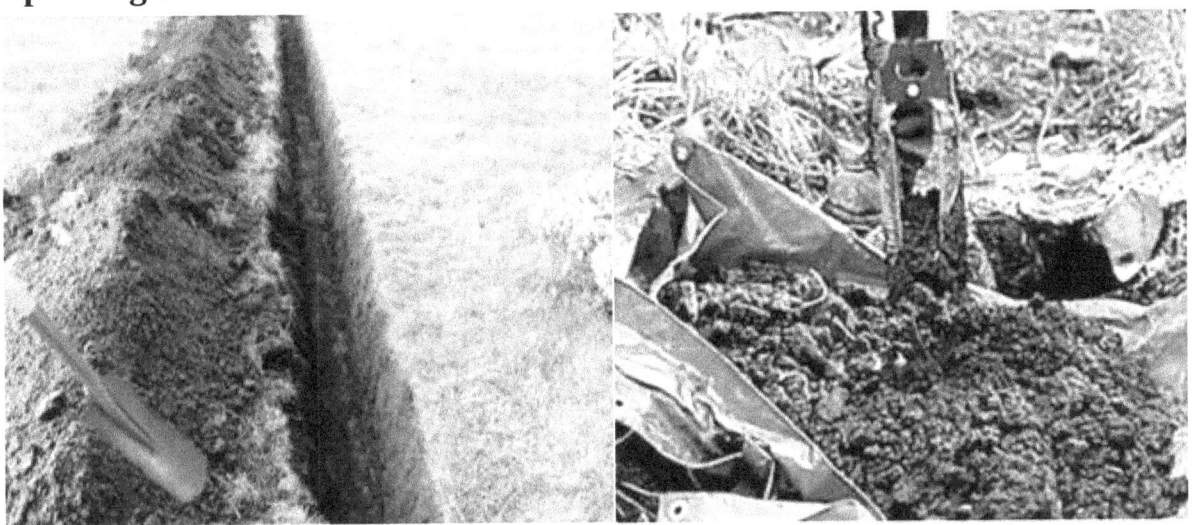

Use a masonry hoe, a round shovel, and a trenching shovel to dig a 12 inches deep trench from the posthole to the wall where the power pocket is. It doesn't matter how wide the trench is, but it is important to make sure it's 12 inches deep to avoid the risk of exposing buried wires in case something or someone digs into the soil.

If your yard has grass or shrubs, put a plastic tarp on one side of the trench to place the soil onto, thereby keeping dirt off the grass or plants and making it easier to clean up after the project. In addition, it will make your work easier when it's time to fill the soil back into the trench.

If your trench comes across a footpath, skip the footpath and continue digging on its other side until your trench connects with the posthole.

## Step 5: Install the conduit

Conduit will help protect the underground wires from moisture. The best type of conduit to use is the grey PVC one.

Lay the conduit on the side of the trench (the one without soil) to measure and cut it to the required length, from the wall to the posthole. Use a tape measure to measure the trench from the inside and use a hacksaw to cut the conduit where necessary (e.g., in places where you have to make turns).

Use 45-degrees or 90-degrees couplings to make the turns and connect the conduits. For example, you will need a coupling to connect the conduit running from the power pack to the conduit laying straight in the trench. You will also need to make a coupling connection between the conduit at the end of the trench and the conduit running up the posthole (it should be 6 inches long, above ground).

Where there is a footpath, drive the PVC conduit under it, using a sledgehammer until it gets to the other side of the trench, then continue laying the pipe until it gets to the posthole.

Use fish tape to thread a 12-gauge UF cable through the conduit. Make sure to leave a long enough wire to get to the power post at the conduit end near the wall, and an extra 9 feet wire at the conduit end in the posthole. Remember to follow the instructions given earlier on using fish tape.

You can make your work easier by feeding the wire through the conduit before making turn connections with couplings. This means if you have 2 conduit pipes and a coupling, you can start by feeding the wire to one of the pipes, to coupling (join them with the wire now at one end of coupling), then thread the wire into the other conduit before connecting it (the conduit) to the coupling conduit. After that, you then lay the conduit in the trench.

## Step 6: Fill the trench

Use a shovel to backfill the soil into the trench to cover the conduit pipe. Make sure the pipes are facing the right direction, i.e., if there is a connection, the coupling is laying at the right angle.

## Step 7: Fill the pole hole

First, mix concrete in a wheelbarrow (a bag at a time) and make sure the mix is stiff, i.e., it should stay in its shape and still jiggle when shoved. This kind of concrete will set faster and support the post.

Using a shovel, pour the concrete into the hole (on top of the gravel). Fill the hole until the concrete is at level with the ground or just a few inches below (this should leave at least 6' of conduit pipe exposed). As you pour the concrete, gently chop it with your shovel to help remove air pockets. Be careful not to pour any concrete on the wires or into the pipe.

Smoothen the top of the concrete with a concrete float.

## Step 8: Install the post

This step depends on two things, i.e., if the post has a lamppost base that needs anchoring with bolts and nuts or not.

If the post does not have a base or manufacturer's instructions on installing it, you will have to push it into the concrete. First, lay the post beside the hole and use your fish tape to thread the wire through the post from the bottom until it protrudes at the top. Measure the post and make a mark at 12 inches from the bottom.

Push the post into the concrete as you twist it until the 12 inches mark is in line with the concrete footing. Although you should be careful, you should also be quick to slide the post into the concrete when it is still wet. Ensure the post is straight, then leave it for half a day to let the concrete harden.

If the lamppost has a basement you need to install before the post, it should have installation instructions. In this case, carefully follow the manufacturer's instructions but remember to be quick so that the concrete doesn't harden before you install the post to its foundation.

## Step 9: Connect the UF wire cable to the light

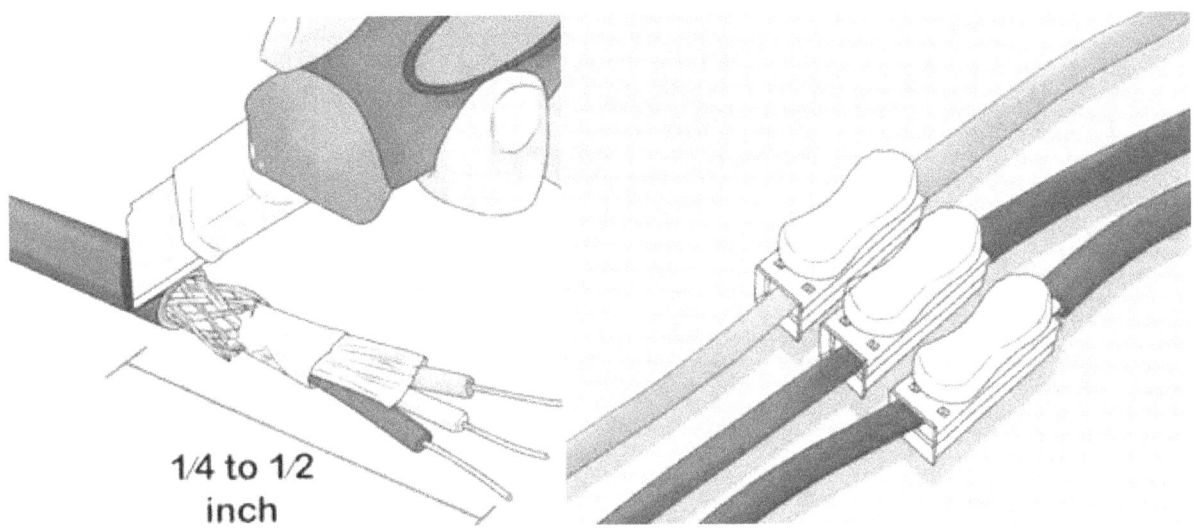

1/4 to 1/2 inch

Use a utility knife to remove 3 inches of the grey sheathing on the UF cable. This will reveal 3 wires, i.e., red/black/brown for hot wire, white for neutral wire, and green or bare copper for the ground/earth wire.

Using a wire stripper, strip off ¼-½ inches of the insulation on each of the 3 wires.

If the lantern comes disassembled, assemble it and remember to follow the manufacturer's instructions since every lantern is different. After that, strip off ¼-½ inches of the insulation on the lantern wires.

Use wire connectors to connect the wires from the ground to those from the lantern. However, this should not be a mix and match kind of connection. Instead, ensure you connect black/red/brown from the post to the black/red/brown from the lantern, white from the post to the white from the lantern, and green/bare copper from the post to the green/bare copper from the lantern.

Tuck the wires into the post and insert the lantern into the post before using a screwdriver to fasten the lantern's screws into place.

## Step 10: Connect the UF wire cable to the power pack

Since the professional who installed the power pack knows the sensor type and sensor wiring method they used, you will have to call them to make this connection, as you do not want to make any mistake that can lead to a failed project and other worse consequences. They will also install the lamppost switch.

## Step 11: Turn on the power

Double-check that all connections are secure, then turn the power back on. Test if your lamppost is working properly by switching "ON" and "OFF."

## 4: Replacing a Light Fixture

Another simple wiring project is "how to replace a light fixture." There are many reasons you would want to change your light fixtures. For example, if it is not bright enough, if it doesn't match your home design or is faulty.

This project is a universal process as almost all lightboxes are of the same size and have 3 connected wires, i.e., hot, neutral, and ground wires. The only thing you need to do is get a new light fixture that will perfectly fit your interior design.

The tools you need for this project are a voltage tester, screwdriver, pliers, wire stripper, a ladder or 2, wire connectors, a new light fixture, and a handy helper.

Before you begin, take your new light fixture and assemble all its parts using the manufacturer's instructions to ensure that everything is as it should be.

### Step 1: Turn off power

Switch off the circuit breaker that provides power to your light fixture. It should have a label indicating that it's the lighting circuit. If you cannot find it, turn off the main switch.

### Step 2: Remove the existing fixture

First, remove any unfastened parts of the light fixture, e.g., glass covers or bulbs.

After that, use a screwdriver to unscrew the canopy (the round cover screwed on the ceiling hiding the wires and the lightbox). If the light fixture you are removing is heavy, let your helper help you manage the weight because the fixture still connects to the circuit wires.

## Step 3: Disconnect the wires

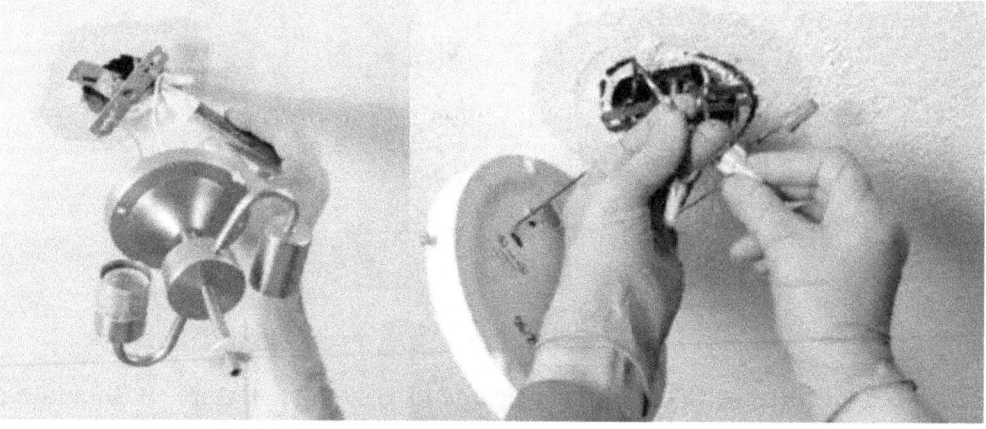

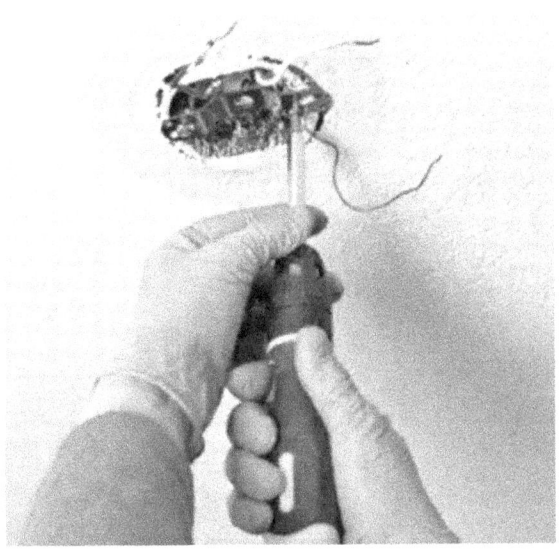

Use a tester to confirm you have no current flowing to the light fixture.

The circuit wires are usually connected to the light fixture wires using wire connectors; hence, you have to remove the wire connectors using a screwdriver to disconnect the wires.

Once you have the 3 wires (black/red in the USA and brown in the UK for hot, white for neutral, and green or bare copper for ground/earth) disconnected from the fixture wires, the light fixture should be held in place by a screw or 2 screwed on the lightbox. Loosen these screws to free the light fixture.

## Step 4: Strip the circuit wires and the new fixture wires

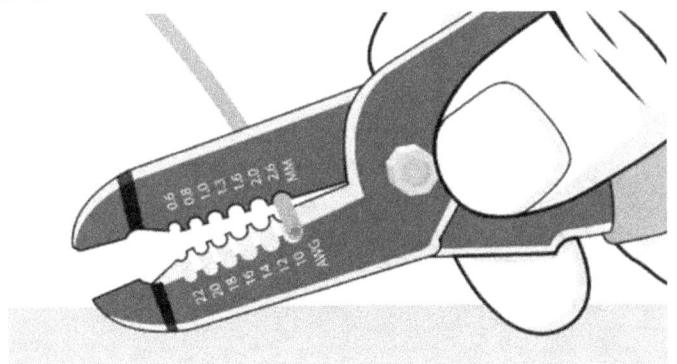

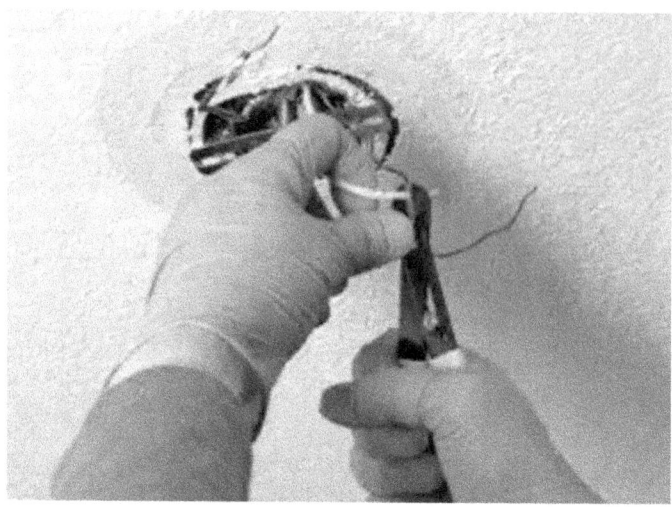

If installing a hanging pendant or a chandelier, figure out how far you want it to hang from the ceiling. However, it should be at least 30 inches up from the table for better light.

Cut the fixture wires to the desired length and add a few extra inches (maybe 6) for making the connection. If the fixture mounts onto the ceiling or wall, 6-8 inches of each wire should be enough.

Strip 1 inch of insulation off the fixture wires using a wire stripper.

Check the circuit wires if they are in good condition. If frayed, cut the exposed part off and strip an inch off the insulation on each wire.

## Step 5: Install the new mounting bracket

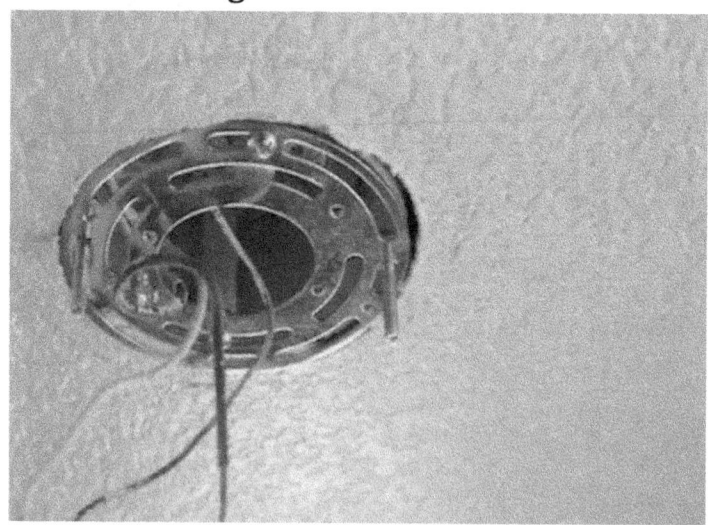

The new fixture should have a bracket plate used as the hardware for securing it. Use the manufacturer's guide to install the bracket on the electrical box since they can vary in shape, e.g., the circular type and the "strap" type with "nipple."

Do not hide the circuit wires in the bracket by weaving them through the bracket.

## Step 6: Connect the wires

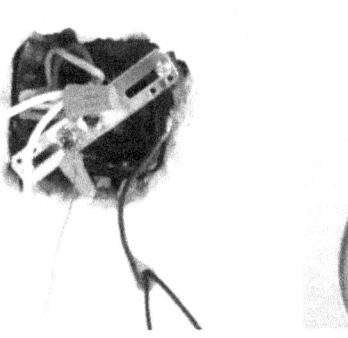 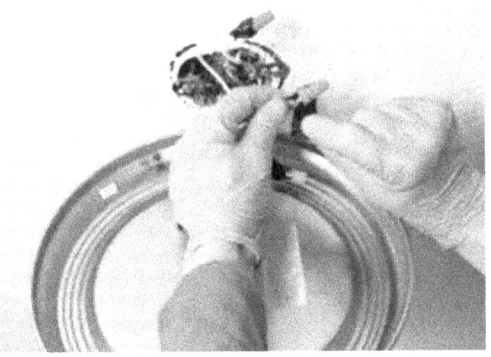

If the new light fixture you are installing is heavy, let your helper support its weight as you connect its wires (that resemble frayed silver) to the exposed circuit wires using wire connectors. Remember to wire correctly by connecting the circuit wires to matching fixture wire, i.e., connect hot to hot, neutral to neutral, and ground to ground.

However, some electrical boxes have a green screw instead of the earth wire. If this is the case, wrap the fixture's neutral wire on the green screw terminal.

Ensure the hot wires do not come into contact with the ground wires/screw by connecting them at different sides of the bracket plate.

## Step 7: Secure the new light fixture

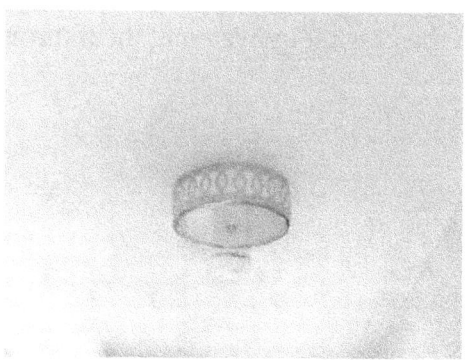

If your light fixture is to sit flush on the ceiling or wall, tuck the wires into the electrical box, then use a screwdriver to install the base of the fixture.

If the light fixture has chains for suspending it on the air, adjust them to the desired height, tuck the wires and any extra chains into the ceiling and install the canopy. After that, use the fixture's screws to mount it on the canopy plate.

## Step 8: Attach bulb and bulb covers

Attach the recommended bulb type for the fixture (check the manufacturer's guide) and use the same guide to attach the bulb covers.

## Step 9: Turn the power back on

Turn on the power at the main switch or circuit breaker and test to see if your light fixture is operating properly by turning it "ON" and "OFF" at the switch.

## 5: Installing an Electric Cooker

A time will come when you need to install a new electric cooker, maybe because the old one is dead, you want to upgrade your kitchen appliances, or for any other reason. When that time comes, there are some legalities you need to meet before you begin this project.

You might be wondering why you cannot just plug your cooker into an outlet and be done with it. Here is why:

Cookers need a lot of power to operate. Hence, they need an individual circuit. If you are replacing an existing cooker, it means your home has a cooker circuit. However, before you install your new one, you need to check if the existing circuit has enough power to provide the required amount of electricity.

Check the amp rating of the cooker (most modern cookers have a 32 amperage rating) and compare it with that of the cooker circuit fuse (in the circuit breaker). If the cooker amperage rating is higher than that of the existing power circuit, you need to have the circuit replaced (to the required one) by a licensed electrician. If your home doesn't have any existing cooker circuit, have a licensed electrician install one.

Since cookers need a lot of power, they must be "supervised" using double pole isolating switches. A double pole isolating switch is a type of switch designed with 2 toggles that help shut off the hot and neutral wires should a short-circuit happen, thereby preventing shock.

If you had a cooker installed before, you probably have this switch somewhere in your kitchen, and you won't need a new one (you will have to confirm if it is properly working using a voltage tester). However, if it is the first time you are installing a cooker in your home, you have to hire a professional to install it.

Remember: don't just go for any switch type because cookers require more energy that's retainable in the neutral wire. The ordinary switch will not protect you or your appliance as it only de-energizes the hot wire.

Also, be mindful when it comes to positioning your cooker. It should not be directly under its control unit as fire can blast onto the circuit's wire, leading to damages and accidents. Also, it should not be far away from the unit as the wires connecting the two are short.

Keep your cooker at least 2 meters to either side of the control unit (depending on the size/design of your kitchen). Ensure no flammable objects, e.g., wallpapers and wood, are near the cooker. Also, if you are installing your cooker in a new home or a renovated kitchen, keep about 300mm between it and other appliances, e.g., fridge or dishwasher.

Every new cooker comes with its cables, but if otherwise (maybe you are installing an old cooker), you will need to buy a cable. Go for a 2.5mm (0.098 inches) heat resistant cable, usually known as "twin and earth" or "2 core and earth," as it has enough thickness for this wiring project. Makes sure the cable has 3 wires, i.e., hot, neutral, and earth, which will vary in color depending on your country's color code (refer to the images on color codes).

There are no other requirements for this project besides the tools in your toolbox, a multi-meter, a cooker, and a cable.

## Step 1: Turn off the power

Start by turning off the power for the cooker's circuit at the circuit breaker and lock the consumer unit or leave a note so that someone doesn't turn it on.

## Step 2: Connect the cable to the cooker

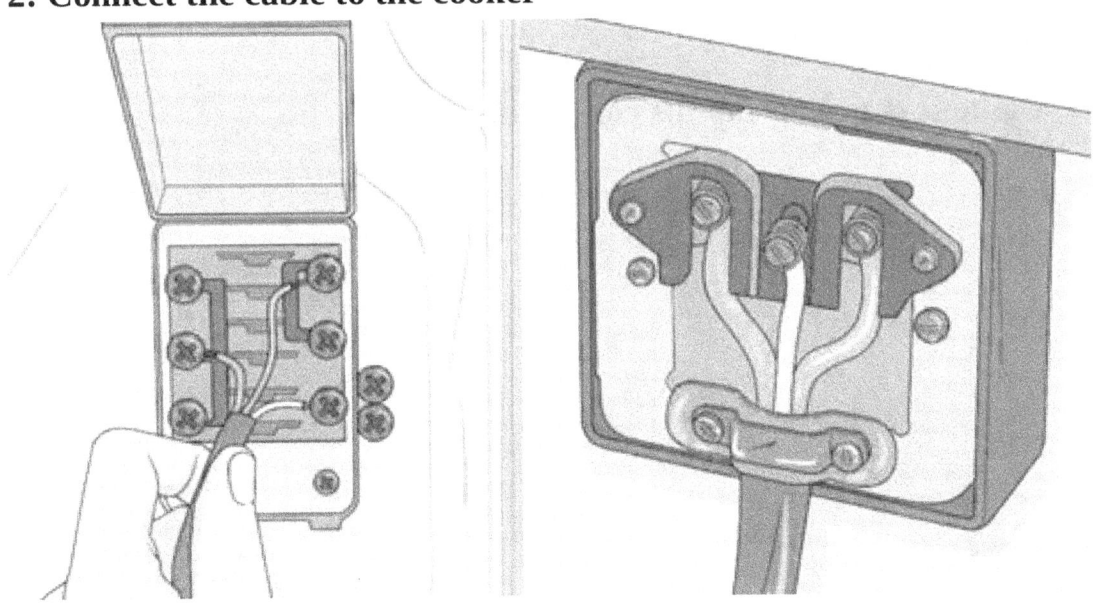

With the cooker away from the wall, look for a small box (usually at the right bottom side) at the back. This box has a panel cover mounted with screws or designed to fit in place without screws.

If it has screws, use a screwdriver to loosen them and if it doesn't have screws, use a flathead screwdriver to pry and lift it out of place. Unscrew the cooker's screw terminals in the box—some cookers have 3 screws, while others have 6.

In cases of 3 terminals, unscrew all of them, i.e., the right, left, and center screws; in case of 6 terminals, only unscrew 3 of them, i.e., bottom right, upper right, and center-left screws. Check which screw terminal is for which wire in the user manual.

Cut the cable sheathing and then strip 0.5 inches of insulation off each wire (the ground wire might be bare copper —hence no need for stripping) and connect them to their right screw terminals (each wire on its terminal). Let the manufacturer's instructions guide and help you identify the terminals and use your country's color code to identify the wire. For example, if you are from the UK, blue means neutral wire, and if you are from the USA, red means live wire).

If you do not have the user manual, attach the hot wire to the right screw, the neutral wire to the left screw, and the earth wire to the center screw for a cooker box with 3 terminals. For the one with 6 terminals, attach the hot wire to the center-left screw, the neutral wire to the top right screw, and the earth wire to the bottom right screw.

After warping the wires onto their right screws, use a screwdriver to secure each into place and pull the wires to confirm they are tight. Make sure you leave no bare wire exposed.

Push the cover plate into place (if it doesn't have screws) or use a screwdriver to drive and tighten its screws. Make sure you gently and neatly tuck in any wire without sheathing that's sticking out of the box.

**Step 3: Connect the cooker to the control unit**

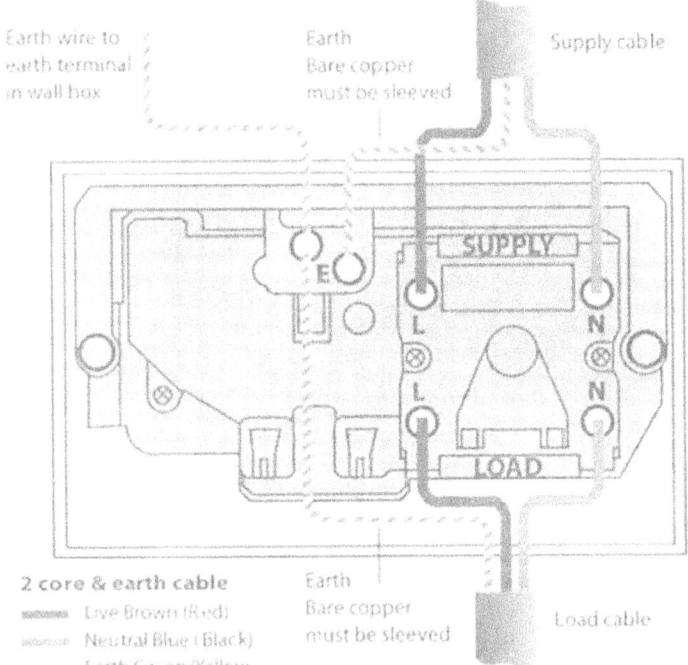

Cooker control switch with socket

Turn off the isolating switch to ensure no electricity flows to the control unit. Use a screwdriver to remove the screws holding the cover plate in place. Before commencing wiring, use a multimeter to test for voltage since cookers use high voltage current, which can be dangerous if you touch it.

Plug in the multimeter leads, i.e., the back and the red, into place and turn on the tester. Set it to AC voltage, then put the tip of the black lead on the control unit's ground terminal (at the center) and that of the red lead on the unit's hot terminal (on the right).

If no current flows into the control unit, the tester's reading will remain at zero. If you get any reading, turn on the isolating switch to de-energize the control unit, then turn the switch back off.

Remove the screws on the control unit terminals (usually 3) to open the slots in which you should fit the wires. Inside these slots, you will see the circuit wires (one on each of the 3 slots).

Strip off ½ inches of insulation from each wire and insert them into their right slot. Insert the live cooker wire into the slot with live control unit wire (the right one), the neutral cooker wire into the slot with neutral control unit wire (the left one), and the ground cooker wire into the slot with ground control unit wire (the center one). Use a screwdriver to tuck these wires into their slots. Ensure that the wires are not frayed and that they properly connect with those in the slot because the cooker won't work well in case of bad wiring.

Place the screws back into the terminals and use a screwdriver to tighten them. Make sure not to drive the screws at an angle (cross-thread) because it can cause a poor connection.

After confirming that you have made all the 3 connections properly, reinstall the cover plate.

## Step 4: Have your work inspected

Since this project involves a lot of electricity, you have to ensure that everything is in order. Have a licensed electrician check your work and certify it. Yes, your work might be perfect, but you might have a hard time insuring or even selling your property in the future without this certificate.

## Step 5: Turn the power back on

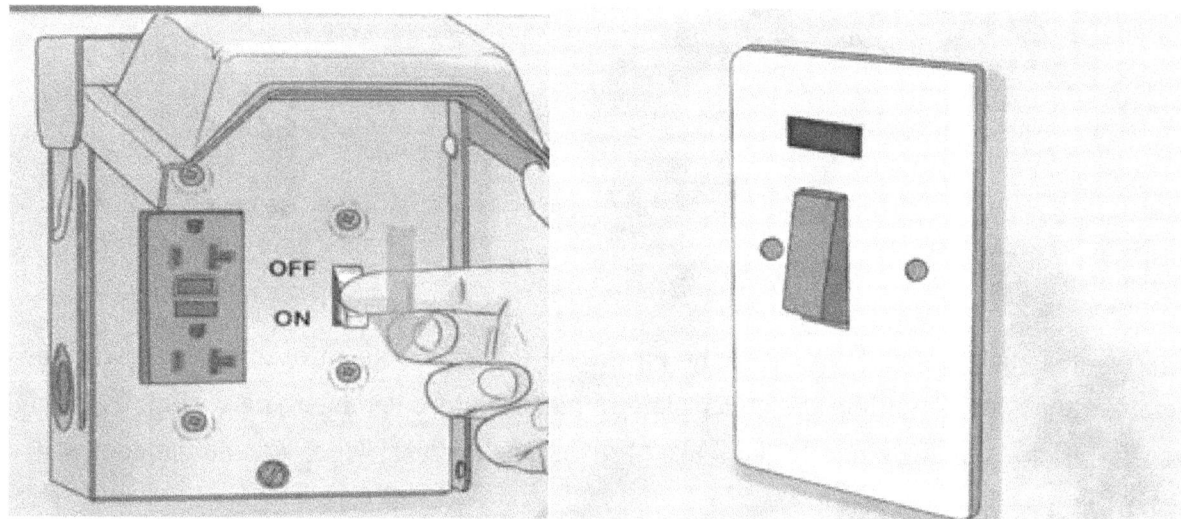

After the electrician has confirmed that the wiring is okay or maybe rectified any mistakes, connect the other parts of the cooker (following the manual) and turn on the power at the circuit breaker and the double pole isolating switch.

## 6: Installing a Doorbell

Another simple wiring project you can try is installing an electric doorbell. If your home doesn't have a doorbell, if it's broken or, if you simply want to install a modernized one, this is a project you do not want to pass on.

There are 3 types of doorbells from which to choose. The choice you make will depend on several factors, like your budget, preferences, and existing wiring system. These types of doorbells are:

## Wired doorbell

As the name suggests, a wired doorbell has hard-wired components to a circuit to draw its power. It is ideal for a home with an existing doorbell wiring system. However, if your home doesn't have this system and you prefer this type of doorbell, you can have a licensed electrician install the circuit system. A wired doorbell system has a chime unit, a transformer, and a push-button unit.

The chime unit is the part of the bell system that produces the "ding dong" sound when someone presses or pushes the button at the door. It has 3 terminals at the center hub labeled "front," "trans," and "rear." The "trans" terminal connects the wire from the transformer, while the "front" terminal connects wires from the front push button. The "rear" terminal connects the wire from the rear push button (if one is connected to the doorbell system). The Chime unit should be strategically placed, perhaps in the living room, the kitchen, or any central location where people in the house can easily hear it from any room. This component is available in different features and dimensions; choose the one that suits you.

The transformer is an important component of the doorbell system. Since most push buttons are installed outside the house, they are likely to experience all sorts of weather and, directly connecting them to 240-V (the basic household circuit voltage) can be dangerous. However, the transformer reduces this voltage from 240v to 12v or less, making the push button safe for people to press. It can be located in any inconspicuous area outside or inside the house, but it's best mounted in an electrical junction box, enclosing the high voltage wires.

The last component is the push button unit (the doorbell part that people press). Pressing it causes the chime unit to produce a sound, notifying you or anyone in the house that someone is at the door.

Push buttons are of 2 types, i.e., one hard-wired to the chime unit through the transformer, and the wireless ones that use batteries as their source of power. Push-button units should always be installed at either side of the door or on the door for easy spotting. Although push-button units come in different designs, colors, and styles, it is a good bet to purchase a weatherproof model that can withstand snow and rain.

Here is how to install a wired doorbell

## Step 1: Turn the power off

On your circuit breaker, turn off the switch providing electricity to the circuit you intend to work on. If you are unsure which one it is, turn off the power to the premises at the main switch.

## Step 2: Remove the existing doorbell system

If you are replacing a doorbell system, you will first have to uninstall the old chime unit, transformer, and push-button unit.

## Step 3: Removing the chime unit

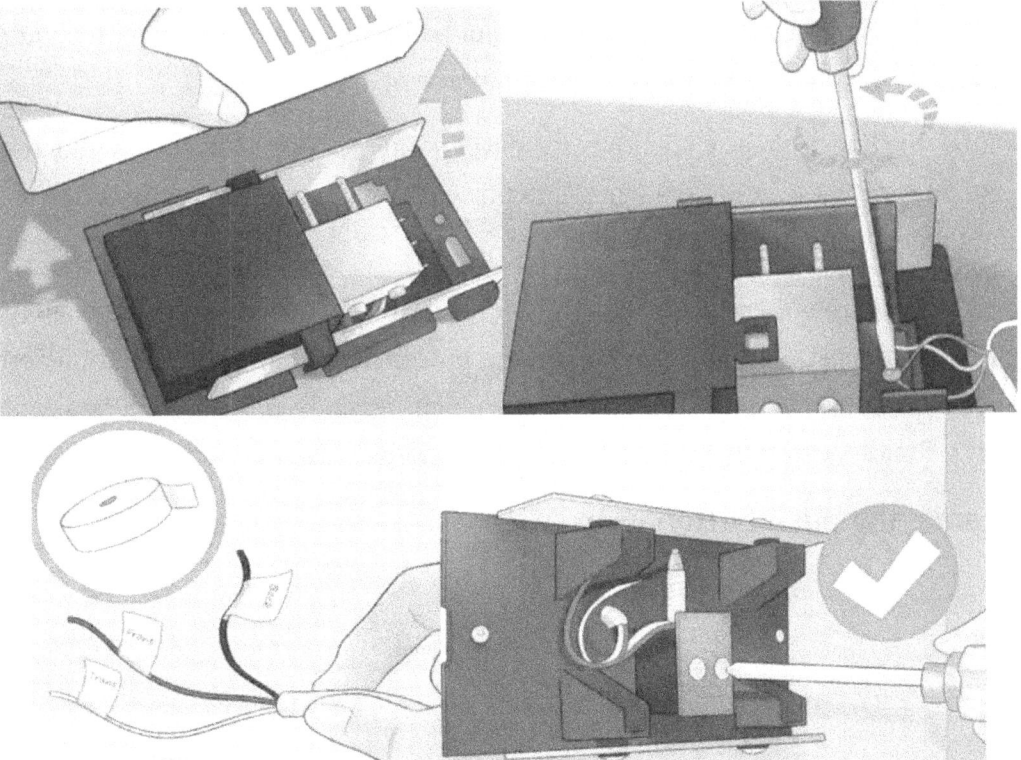

To remove the chime unit, remove its cover plate by prying and lifting it out of place using a screwdriver. Inside this unit, you will find 2 or 3 terminals connected to wires. Wrap each wire with electrical tape and mark them to indicate the chime unit terminal they will connect to, i.e., "front," "trans," or "rear." Use a screwdriver to loosen the screws hence disconnecting the wires.

After the wires are free, use your screwdriver to loosen the screws on the sides of the chime unit, hence unmounting it. Make sure to hold the unit with one hand as you remove the last screw to prevent it from falling on the floor (an occurrence that can damage the floor).

## Step 4: Removing the transformer

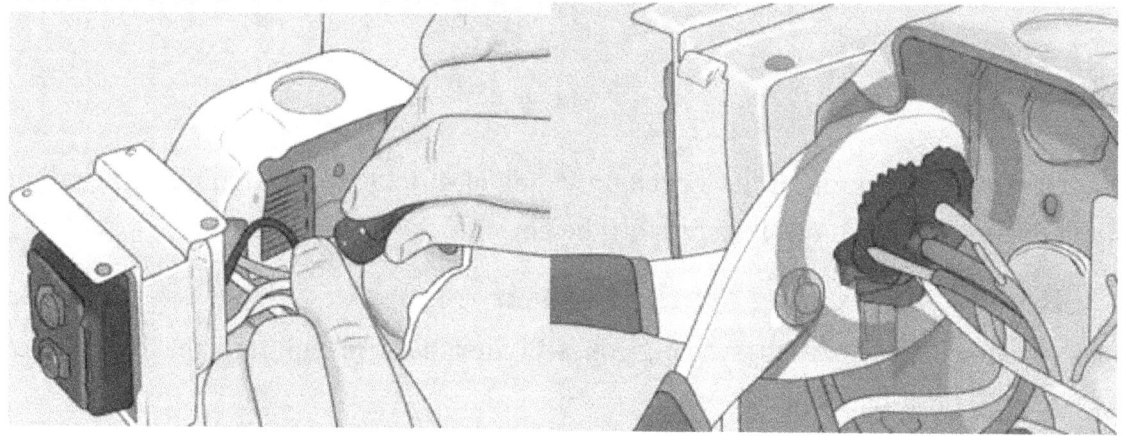

To remove the transformer, you first have to locate it. There is no specific area for installing this component, but it might be on the outer wall of your basement, attic, or garage. It should be in an electrical junction box in proximity to the push button unit.

After locating the junction box, remove its cover plate so you can access the transformer. Unscrew the 2 terminals securing the wires into place (the wires leading to the chime unit and the wires going to the push button unit).

The transformer wires are connected to the circuit wires by wire connectors. Untwist these connectors to disconnect the transformer from the power wires. To detach the transformer from the electrical box, use a screwdriver or a wrench to loosen the large screws or bolts (respectively). Finally, carefully pull the transformer and its wires out of the electrical box.

## Step 5: Removing push button unit

To remove the push button unit, loosen the screws securing the push button on the wall. If these screws are hidden under a cover plate, use a flathead screwdriver or a utility knife to pry and lift the plate to expose them.

After detaching the push button from the wall, gently pull it to expose the wires attached to it. Use a screwdriver to loosen the screws holding the wires into place, thereby disconnecting them from the push button, then gently pull the push button and wires away from each other.

## Step 6: Install the new wired doorbell system

To install the new doorbell system, you have to work on each component at a time. So, let's start with installing the new transformer through the following steps:

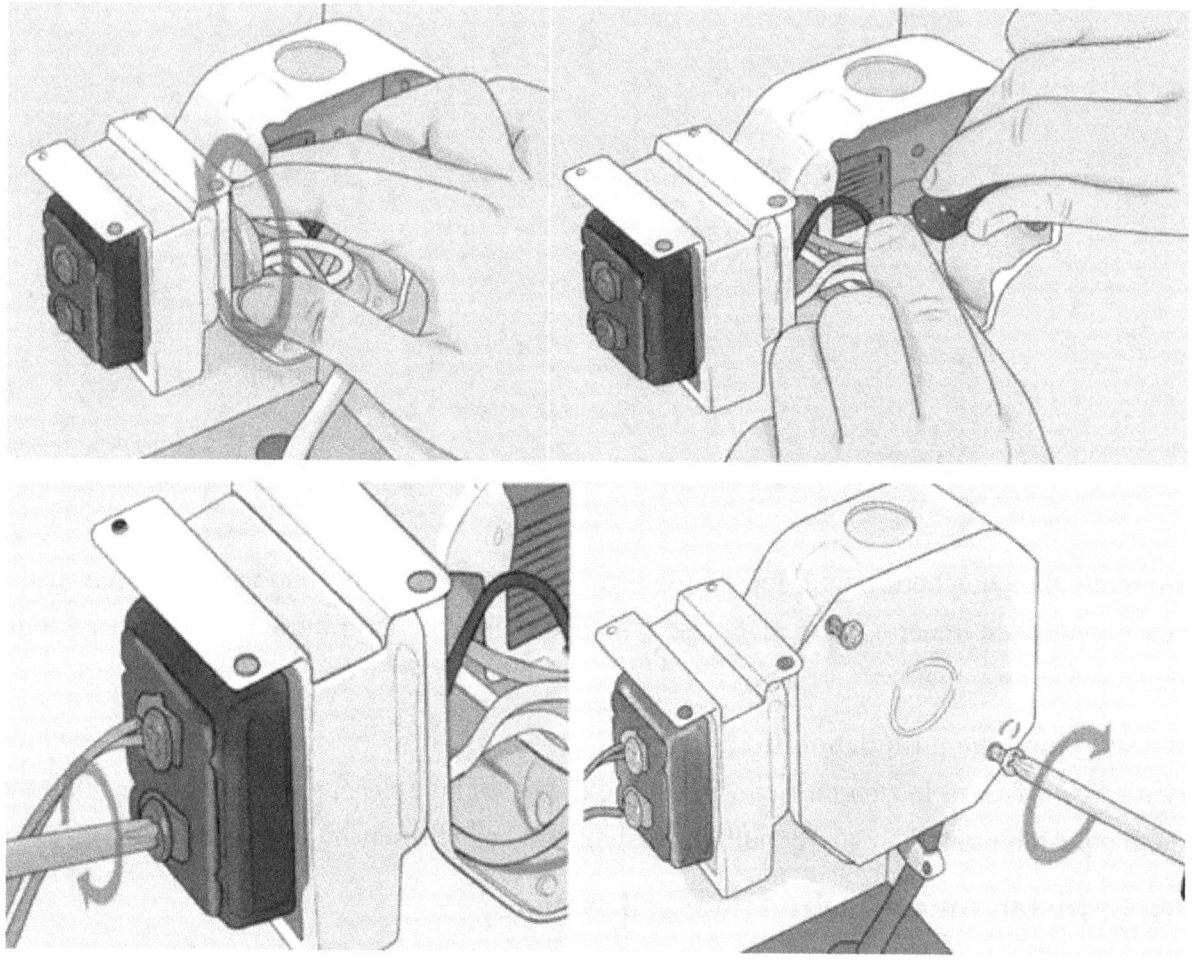

- Attaching the transformer to the junction box: Remove the cover plate and fit the transformer in its space in the box. If the transformer came with bolts, use a wrench and if it came with screws, use a screwdriver to secure it in place.

- Connect the circuit wires to the transformer wires: if the circuit's bare wires are frayed, cut them off and strip off about 0.5 inches of insulation from each wire. Also, strip off 0.5 inches of insulation from the transformer's wires using a wire stripper. Match the wires, i.e., the hot wire from the circuit to the hot wire from the transformer, neutral to neutral, and ground to ground, depending on the wire color code in your country. For example, match the brown circuit wire to the brown transformer wire in the UK (live wires) or red/black circuit wire to red/black transformer wire in the USA (live wire). Twist each same wires together, then use a wire connector to secure each pair in place.

- Attach the chime wire and the push button unit wire to the transformer. Wrap the chime wire and the push button wire on either of the 2 screw terminals on the transformer. Use a screwdriver to tighten the screws hence securing the wires in place.

- Reattach the junction box cover plate by snapping it in or mounting it with its screws.

After installing the transformer, move to the chime unit location and install it through the following steps:

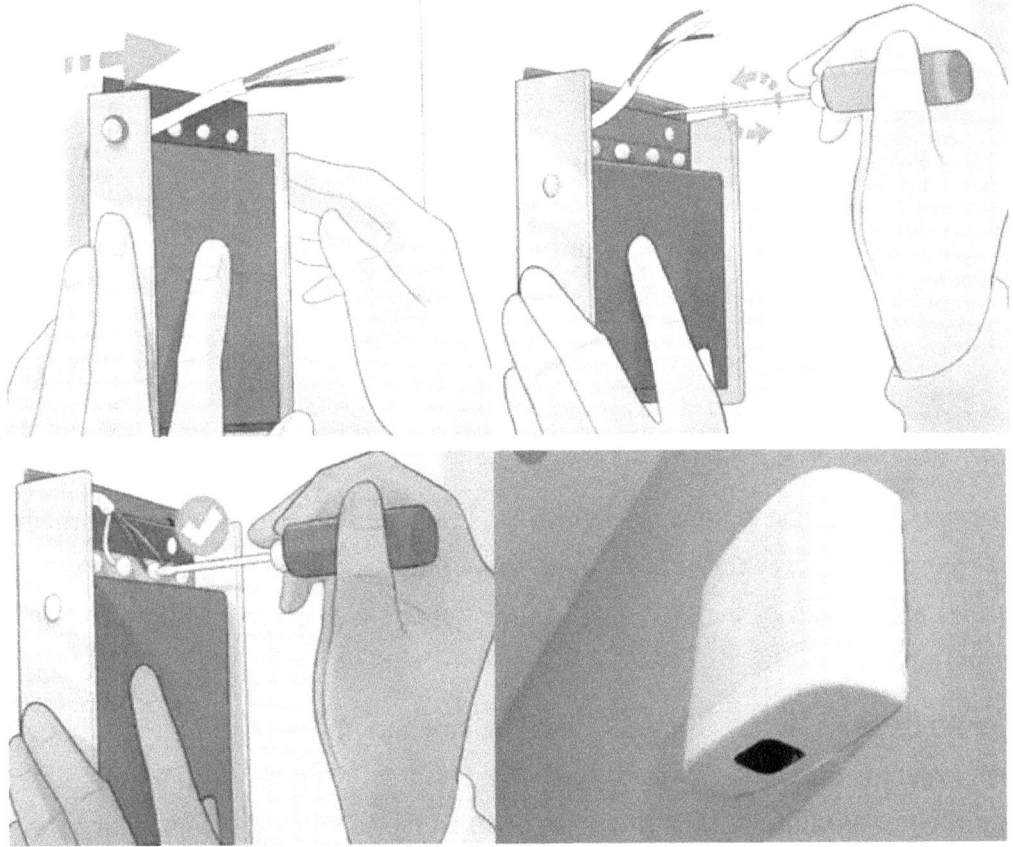

- Identify the wires you or the electrician marked as you thread them into their appropriate holes at the back of the unit, i.e., each wire should go through the hole leading to its right terminal. Be careful not to damage the marks on the wires.

- Mount the chime unit on the wall: Place the unit on the wall, and if it's of the same size as the old unit, use the holes on the wall to secure the chime unit in place using its screws. If this is the first time you're installing a doorbell system in your home, or if the new doorbell is of a different size from the old one, place the screws on their holes in the unit, then use a power drill to drive the screws into the wall (this step will create new holes on the wall).

- Connect the wires to their terminals: If the wires are not in good condition, cut the old bare wire and use a wire stripper to strip off 0.5 inches of insulation from each wire. Use a screwdriver to secure the front push button wire to the "front" terminal, the transformer wire to the "trans" terminal, and if there is a rear push button wire, connect it to the "rear" terminal. Ensure you have no bare wire exposed and all connections on the terminals are tight.

- Fit in the chime unit cover plate by screwing it into place. Check the manufacturer's instructions to make sure that you place it right.

Lastly, install the push button unit through the following steps:

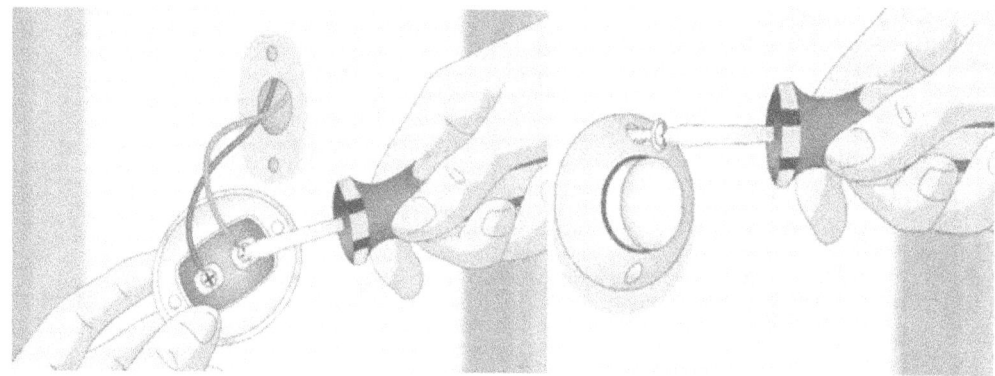

- Connect the push button to the wires: Use your screwdriver to secure the 2 wires emerging from the exterior wall near your door under the 2 screws (each wire on its screw) at the back of the push button unit. Ensure the screws are tight to hold the wires firmly in place.

- Secure the push button in place: Push the unit into the wall to hide all the wires. If the new button is the same size as the old one, use a screwdriver to fit and tighten its screws into the holes on the wall. If there are no existing holes on the wall, place the screws on the unit holes and use a power drill to drive them into the wall.

- Attach the cover plate: If your push button has a cover plate, use the manufacturer's instructions to install it.

## Step 7: Turn the power back on

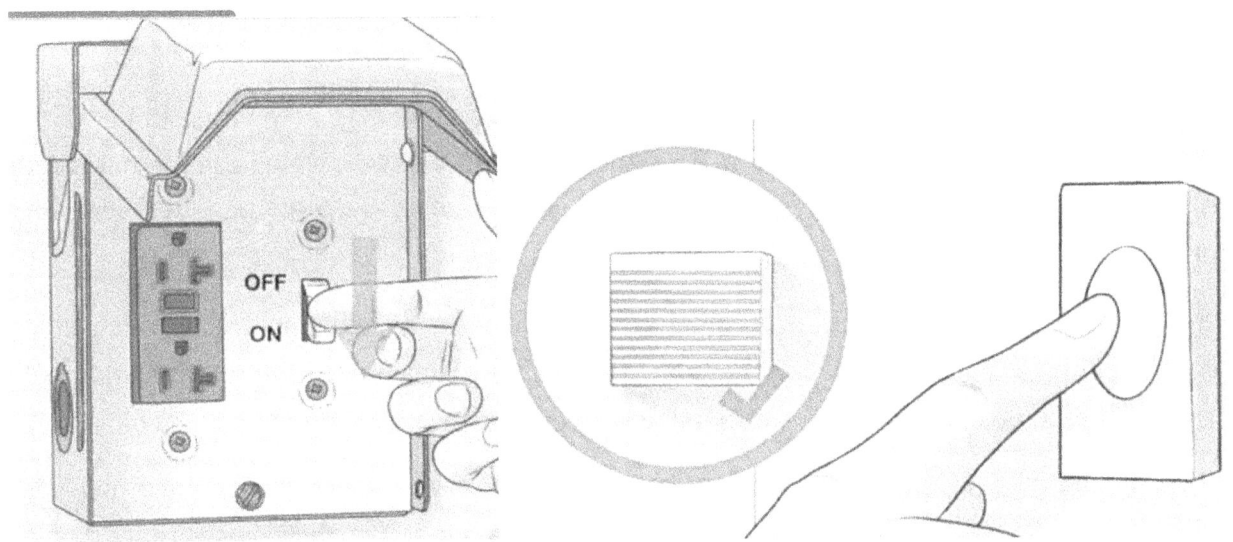

After connecting all the components on your doorbell system, turn the power back on and press the push button to test if the system is functioning properly (the chime unit should produce a sound every time you press the button).

## Wireless Doorbell

The second type of doorbell is the wireless type. Although these types are easier to install than the wired type, they are more expensive. Also, they use battery power, which can run out without your knowledge, thereby failing to do their purpose (ring when someone presses the doorbell button).

Regardless, a wireless doorbell is ideal if you do not have an existing wiring circuit for a doorbell or if you like the idea of having a moveable chime unit. Unlike wired doorbells, wireless doorbells do not have transformers; they only have the push button and the chime unit.

When purchasing a wireless doorbell system, make sure to go for the one that will provide a signal if you move the chime unit to the furthest area in your home (they range from 700-1200 feet signal range). Also, if you are looking to install more than one chime unit or push buttons, explain your specifications to the seller to ensure you can get a recommendation for the best type based on your needs.

Here's how to install a wireless doorbell:

### Step 1: Synchronize the push button and the chime unit

The chime unit for a wireless doorbell has two buttons, i.e., the chime control button that allows you to choose a mode, e.g., motion detector mode or chime mode, and the chime tune button that allows you to set a tune.

The doorbell package comes with batteries for powering the push button and the chime unit (although some chime units come with a plug; hence should be charged through an outlet)- make sure to insert the batteries in place.

To synchronize the two components, press the control button, set it on chime mode, then press the push button. Having synchronized your doorbell system, you can select a preferred tune out of the few provided by pressing the chime tune button and then quickly releasing it once your favorite tune starts playing. You can set the volume to your preferred level.

### Step 2: Install the push button

First, find the perfect location for your push button, i.e., an easy-to-spot and reach location. If you have a door made of vinyl, wood, or aluminum, it is the perfect place to install your push button, as these materials do not interfere with signals. If your door is metallic, you will have to install a wooden shim (using screws and a power drill) before installing the button. Metals interfere with signals; your chime unit will not ring if this happens.

If you want to install the push button on a brick or masonry wall, attach a plastic material to the surface (using screws and a power drill) to help make work easier once it's time to mount the button. Also, install the button at most 4 feet above the ground, which is the highest access point of a wheelchair.

Use adhesive or screws (depending on the one provided by the manufacturer) to mount and secure the button in place. Follow the manufacturer's instructions to install the push button. If you are using adhesive, make sure to clean the area you will attach the button with clean, damp material to make the adhesive stick better.

### Step 3: Installing the chime unit/receiver

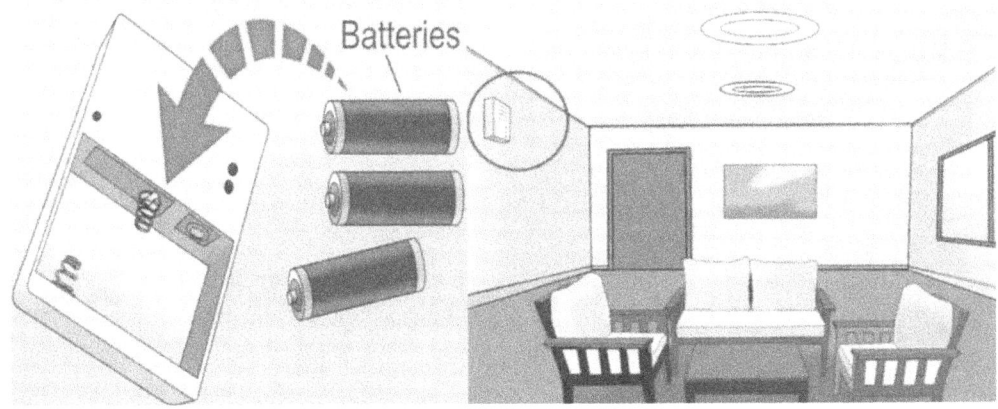

You might have a good doorbell system and a well-installed push button, but if the chime unit is out of the maximum signal range, your doorbell will not ring. If your house is large, buy a doorbell extender (usually another chime unit) as it "forwards" signal to a chime far from the push button, making it possible to carry your chime unit to the furthest end of your premises.

You can place or install the chime unit anywhere as long as it is within the signal range and it won't get in your way. Mount it on a wall, place it on a cupboard, or anywhere recommended by the manufacturer. However, if it has a plug, and you are planning to install it on a wall using screws or adhesive, make sure to attach it near a socket outlet.

**Step 4: Test your doorbell system**

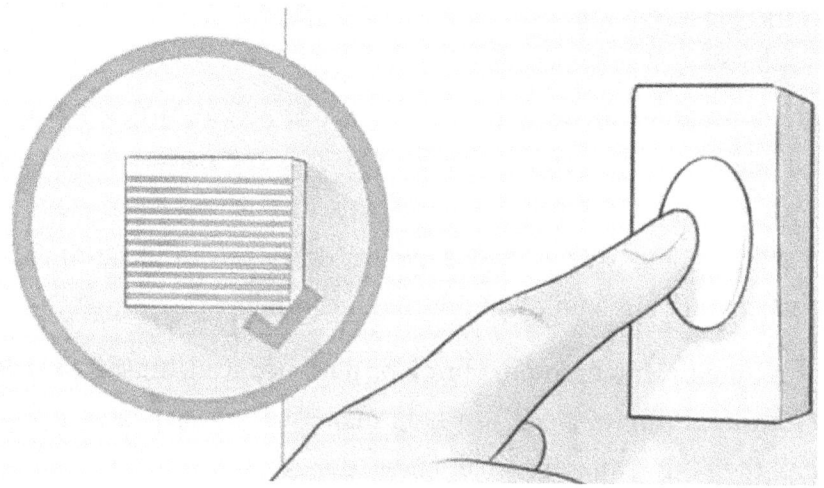

Your wireless doorbell-wiring project is complete. Easy right? Test if your doorbell is working properly by pressing the push button. If it doesn't ring, go through all the steps to see if you skipped any (especially syncing, inserting the batteries, and the receiver placement).

If a nearby push button (e.g., when someone presses your neighbor's push-button, your chime unit rings) activates your chime unit, change your doorbell frequency or channel. Use the manufacturer's instructions to guide you on how to do it.

## 7: Installing Security System

A few years ago, installing a security system used to be a complex process where the system's wiring had to be done simultaneously with the home wiring (during construction). If not, people had to incur additional costs to install a security system. Fortunately, today, we have many modern wireless security systems that are pocket-friendly and easy to install.

A security system's main job is to monitor your home's entry points, e.g., windows and doors, by detecting motion and sounding an alarm. It also monitors and detects water leaks, fires, or any other emergency. Good security systems should be easy to arm and disarm.

Choose a wireless security system that satisfies your needs. For example, do you want it to connect to your phone or want authorities to monitor it? Do you want a system with cameras or not?

It is unnecessary to have your system monitored by authorities if you are working with a tight budget. In such a case, buy a security that will sound the alarm and notify you on your phone. However, this might not be as reliable as the monitored system because you won't get notifications when someone trips it or your phone's battery runs out.

The components of a security system vary based on the manufacturing company, but every system should have sensors, motion detectors, a speaker, and a main panel with a keyboard.

There are no specific steps on installing security systems because each system comes with its instructions. However, some of the tips you should follow during this project are:

## Program the codes

Before you program the codes for the alarm system, you will have to set up mode on the keyboard by entering 6666#. This action will prompt a voice in the unit to say, "Please enter instructions."

Having set up the mode, the system's manual will provide you with the codes for programming different components. Make sure to insert the components' batteries (if not inbuilt) before entering the codes on the keyboard. The system voice will always notify you once programming is complete. You will enter your new password for your security system during this stage. Make sure it is an easy-to-remember password.

Anytime you want to change a setting or program another sensor to the system, you need to set up the mode first. Also, all the sensors have numbers allocated during the coding. You need to number them accordingly for easy installation and know which sensor is triggered when the alarm goes off.

## Installing the main panel

The best place to install the panel is on the interior wall of your house, near your main door. This will enable you to disarm the alarm easily when you get home. However, the exact positioning will depend on its power source, i.e., make sure to install it near an outlet if it has a plug. Depending on the manufacturer's instructions, mount the panel on the wall using adhesive, screws, or small nails.

Always activate the motion sensors before leaving the house or going to bed. The other sensors, e.g., fire or flood, should always be on.

## Installing the sensors

There are different types of sensors you can buy: motion sensors, fires sensors, flood sensors, door and window sensors, etc. Sensors (which come in two parts each) are installed at vulnerable places in your home, like doors and windows, or in the kitchen (fire sensors), to communicate with the alarm system in case of a security breach. The two parts of the sensor are the transmitter and the magnet.

The transmitter is usually bigger than the magnet. Its main role is to transmit a signal when the circuit is broken; on the other hand, the magnet completes the circuit. For example, when a door or a window is opened, the circuit breaks, an action that makes the transmitter send a signal to the panel.

When installing the two parts, ensure they are close to each other (almost touching when the door or window is closed). For example, you can install the sensor on the window frame or door jamb and the magnet on the opposite door/window edge. Before using the provided screws or adhesive to mount them in place, make sure that they are even and lined up.

## Installing the sounder/speakers

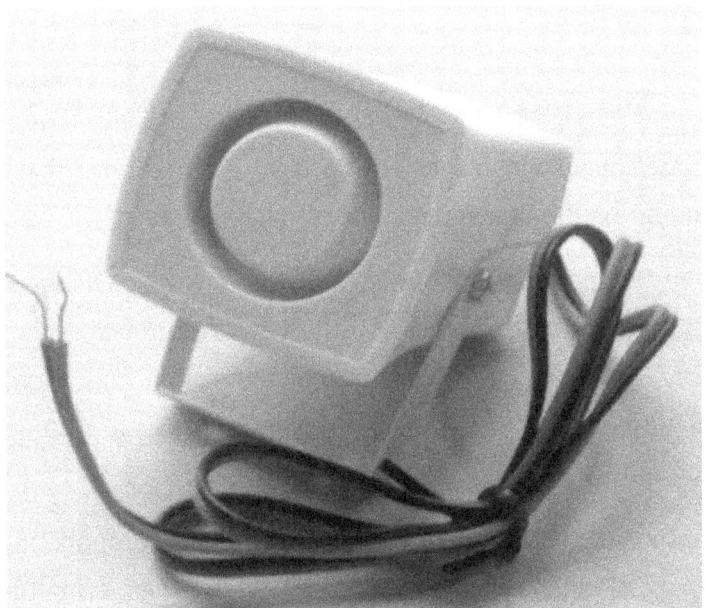

Find a good place where anyone in the house can hear the alarm sound anytime the sensors are triggered. However, make sure to mount it at least 2.4 meters from the ground or the stated height by the manufacturer.

Most speakers have mounting backplates that are installed on the wall. Use the manufacturer's instructions to install it on your chosen location, with the correct side and right side forward and up, respectively. Install the protective box, if any, but do not install the speaker yet.

## Wiring the system

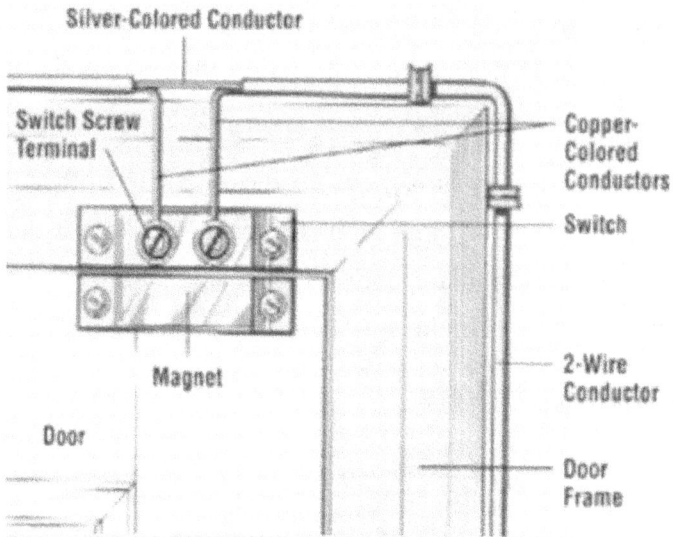

Everything security system package comes with a thin, 2 wired cord that is almost transparent, making it ideal for running on the house wall, doors, or windows.

Strip off 0.75 inches of insulation from one end of the 2 wires. On the furthest transmitter or magnet from the speaker, loop the bare wires on the screw terminals (each wire on its terminal). Without cutting the cord, extend it to the next part of the system, i.e., if you started your wiring on a transmitter, the next part should be the magnet and vice versa. Slip apart the two wires and only cut the copper-colored wire to create 2 ends.

Strip off the insulation from these 2 ends and loop them on the components' terminals. The other wire (silver coated) will not connect to this or any other sensor in this system. It will run alongside the copper-colored one, connecting only on one part of the furthest sensor and one terminal on the speaker.

Continue with this connection on every sensor until you get to the speaker.

Connect the speaker to a battery pack recommended by the manufacturer using the battery wire. Follow the manufacturer's instructions to connect the positive and negative wires to their right terminals.

**Connect the system to your phone**

After connecting all the components as instructed by the manufacturer, use the system's manual to download the best phone app and connect it with the security system.

**Test the security system**

Test if your security system is working properly by opening the door or a window; the speaker should ring, and you should get a notification on your phone. If that is not the case, carefully go through your manual to double-check for mistakes.

## 8: Replacing Shower Isolator Switch

Shower isolator switches, usually 45 amp, tend to malfunction because of the following reasons:

- Turning it off frequently: The main purpose of the isolator switch is to cut off the power supply to your shower when working on the bathroom plumbing system. Do you turn it off every time after taking a shower? If you do, stop as it might be the reason you have many cases of damaged shower isolator switches. Leaving the switch on doesn't consume any power until you turn on the shower; hence, frequently turning it off doesn't save electricity; it only leads to malfunctioning.

- Poor quality: If you choose a low-quality isolator switch, it is likely to stop working sooner than you think. If one type of isolator switch is way cheaper than the others, think twice before buying it as it might end up costing you a lot. To ensure you buy the best quality, consult an electrician and ask for a recommendation on the best type.

- Overheating: If you keep replacing your isolator switch, even though you buy the best quality in the market and keep it on all the time, overheating could be the main culprit. Shower isolator switches tend to overheat when there are improper wire connections, a situation that forces the switches to strain hence, ignition.

A shower isolator switch is in the bathroom on the ceiling with an on and off pull cord or on the outer wall of the bathroom (it doesn't have a cord).

Here are the steps you need to follow to replace your shower isolator switch:

## Step 1: Turn off the power

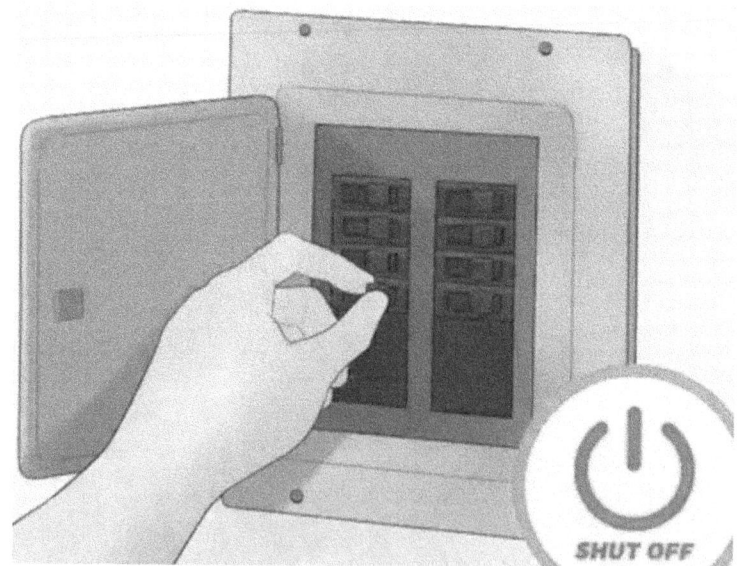

Showers have their own circuit since they consume a lot of power. Locate the circuit breaker leading to the isolator switch and turn it off. If you are unsure which circuit to turn off, switch off the main switch. The red or orange indicator on the switch should not glow.

## Step 2: Remove the old switch cover

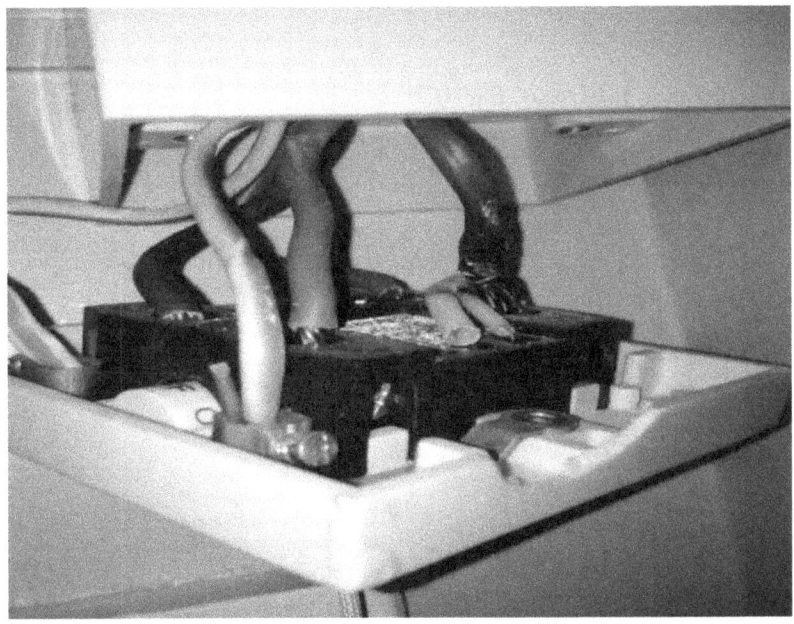

Use a screwdriver to remove the cover plate on the isolator switch. Use a voltage tester to double-check if there is current by placing the tip of the tester on the screw terminals on the switch.

## Step 3: Remove the switch

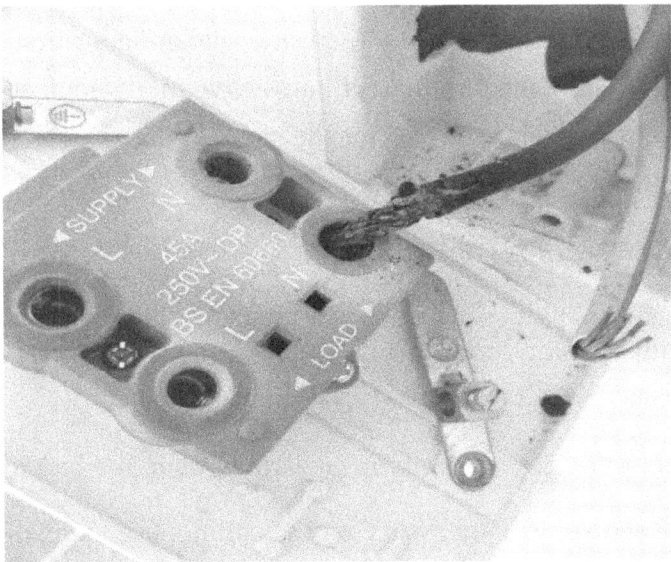

Before removing the switch, note where each wire goes. This kind of switch has 4 terminals: 2

terminals for the incoming live and neutral wires, and 2 terminals for the outgoing live and neutral wires.

Mark them with tape to ensure you don't interchange them when connecting the new switch. Use a screwdriver to loosen the screws on all the terminals and gently pull the wires to free them from the socket.

**Step 4: Connect the new isolator switch**

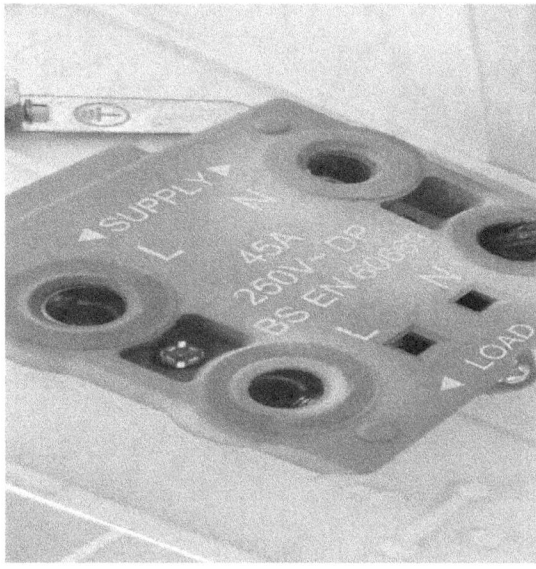

First, check the condition of the circuit wires. If melted or frayed, you will have to cut off the damaged part and strip off the insulation to expose new bare wire. Remember not to remove the mark you put.

On the switch, you will find 4 terminals, 2 on each side, and an indication of whether they are for incoming or outgoing current. Each terminal will also have marks indicating the wire it ought to connect to, i.e., "L" for the live wire and "N" for the neutral wire.

Connect the wires you marked as incoming to their right terminals and do the same for the outgoing. Since the wires are a bit stiffer and thicker than those used to connect single-pole switches for light fixtures ($4mm^2$, $6mm^2$, or $10mm^2$ depending on the power rating for the shower), use a more firm screwdriver to secure them. Ensure that the screws are tight, as isolator switches take much strain.

Although unlikely, if your circuit system offers an earth wire, make sure to connect it.

If you are installing the shower isolator switch on the ceiling, use the manufacturer's instructions to guide to fix the pull cord.

# Finish up

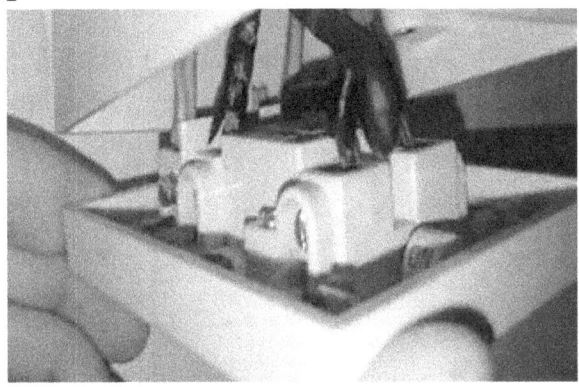

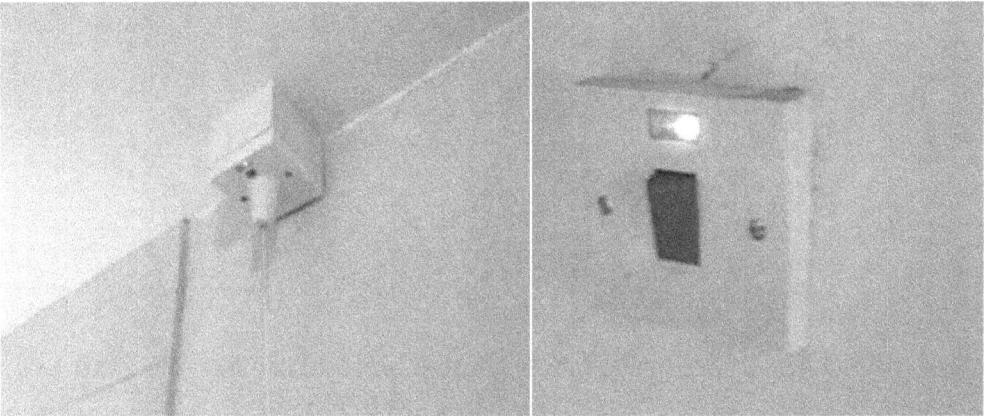

Mount the switch cover plate using the provided screws and a screwdriver. Make sure the indicator neon light is constant.

Turn the power back on, turn the switch on and test if the shower isolator switch is working, i.e., when you turn on the shower, hot water should flow if other parts of the shower are in good condition.

## 9: Installing a 4 Prong Dryer Cord

Clothes dryers are sold without a connection cord because the type required depends on the type of wall outlet, i.e., - if a 3- slot outlet or a four-slot outlet. Most modern houses will have 4-slot outlets; hence, we will focus on how to install a 4-prong dryer cord.

### Step 1: Open the connection box

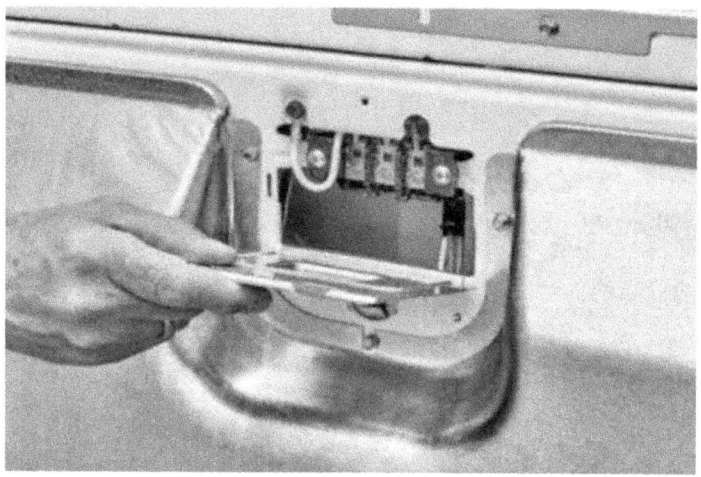

On the back of the electric dryer, locate the connection box and use a screwdriver to unmount the cover plate. The box is located close to a hole of around 0.75 inches in diameter. Doing this will expose 4-wire terminals.

### Step 2: Connect the cord to the dryer

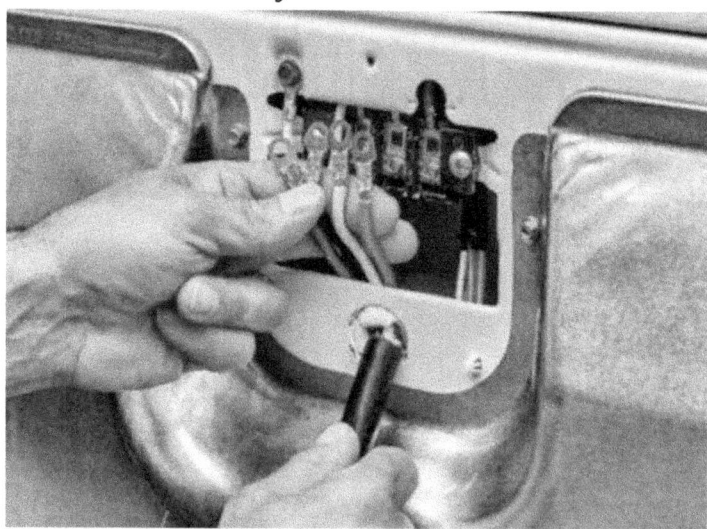

Start by inserting the cord into the hole until it emerges in the connection box.

Strip off about 0.75 inches of sheathing from the cord and the insulation on each wire in the cord.

Wrap the ground wire on the green ground terminal and tighten the screw to secure the wire into place.

On the center terminal, connect the neutral wire, and on the remaining terminals, i.e., the left and the right, connect the two hot wires, each on its terminal. Make sure all the wires are tight and secure on their terminals.

## Step 3: Secure the cord

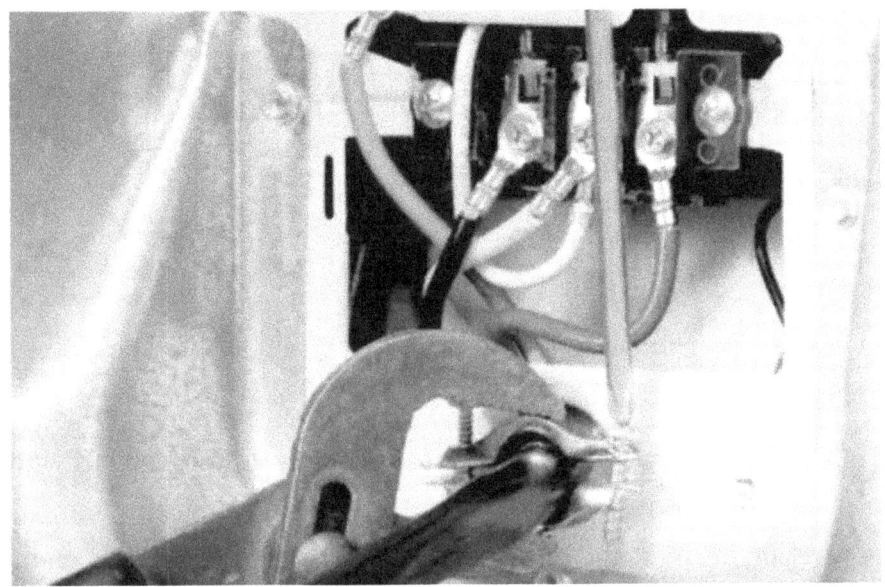

The cord package has a fitting, with bottom and top halves used to relieve strain on the cord. Unscrew the nuts or screws on the fitting to separate it into two. Insert the halves into the entry hole with one-half on top of the cord and the other under the cord. Once they are secure the hole, use a screwdriver to reinstall the screws. To secure the cord, make sure the fitting screws are tight. However, you have to be careful not to pinch the sheathing or deform the wires.

## Step 4: Finish-up

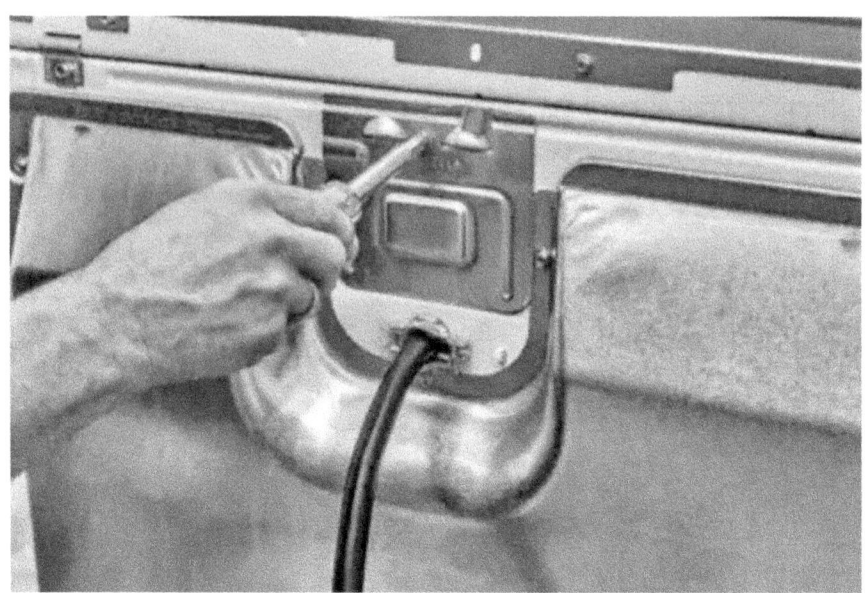

Reattach the cover plate, plug in the cord to a 4-slot socket and turn on the clothes dryer to test if it is working properly.

If your home has a three-slot dryer outlet, it is cheaper to buy a 3-prong cord with three wires (2 hot and 1 neutral) in it than to replace the socket outlet. When attaching the cord to the dryer, connect the live wires on the right and left terminals and the neutral terminal on the ground metal bar in the box (this should leave the neutral terminal unconnected).

## 10: Installing an Icemaker

This project is perfect for you if your refrigerator doesn't have an ice maker but has its configuration or if the old ice maker is not functioning. You will need an ice maker kit with a manufacturer's guide and your toolbox for this project.

Follow the following steps to install a new icemaker:

### Step 1: Unplug the appliance and remove the covers in the freezer wall

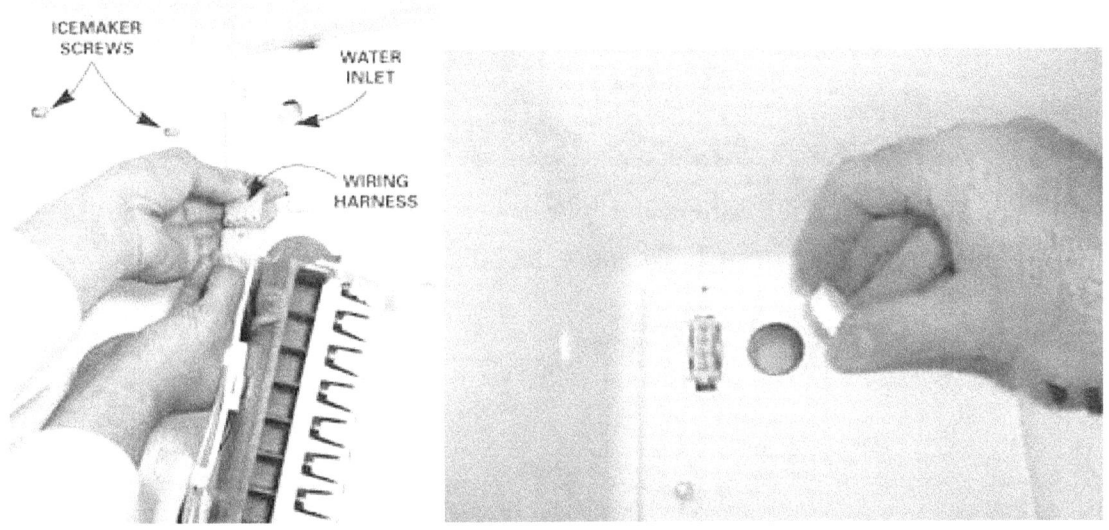

Unplug your refrigerator from the connected socket outlet and remove any shelves in the freezer component.

Use your utility knife or a flathead screwdriver to pry and remove the plastic plugs at the sides and back of the freezer. The exposed holes on the sides are for attaching the icemaker; the round plug at the back connects the water fill tube while the cap at the back acts as the cover plate for the wires.

### Step 2: Uncover the water inlet opening

You will find a sticker marked for the water inlet opening at the back of the freezer. Use a knife to cut the "X" marking on the sticker and expose the hole by removing the foam insulating tube.

## Step 3: Attach the icemaker wires

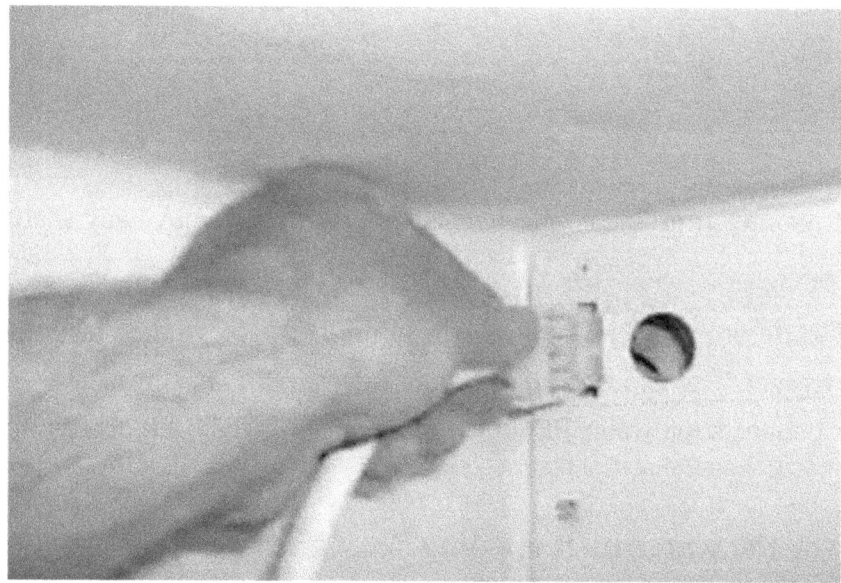

Remove the large cap on the back of the freezer to expose its harness (usually a plug-in). Connect the freezer's harness to that of the icemaker by snapping together. Secure the cap /cover plate back into place.

Follow the manufacturer's guide to identify the different parts in the icemaker kit.

## Step 4: Install the icemaker

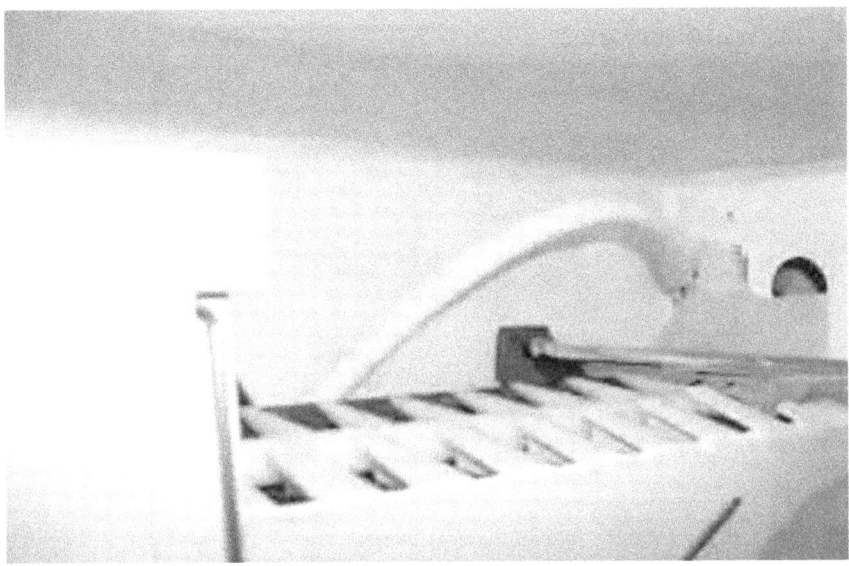

In the holes at the sides of the freezer, partially secure the mounting screws. Place the icemaker on these screws, then tighten them to secure it in place.

Find the leveling bracket (usually L-shaped) and connect it to the ice maker. Make sure that the ice maker is level before tightening the screws.

Place the icemaker tray on the icemaker, then lower it down using the on-off lever.

## Step 5: Install the water fill tube

Into the hole you opened in the back of the freezer, insert the water fill tube that extends from the inlet opening into the fill cup (in the icemaker).

With the "U" shaped end facing up, rotate the tube ¼ turn to the left, press it, then rotate it back to the starting position to install it into place. Place the insulating tube on top of the fill tube and press it.

## Step 6: Install the water tubing

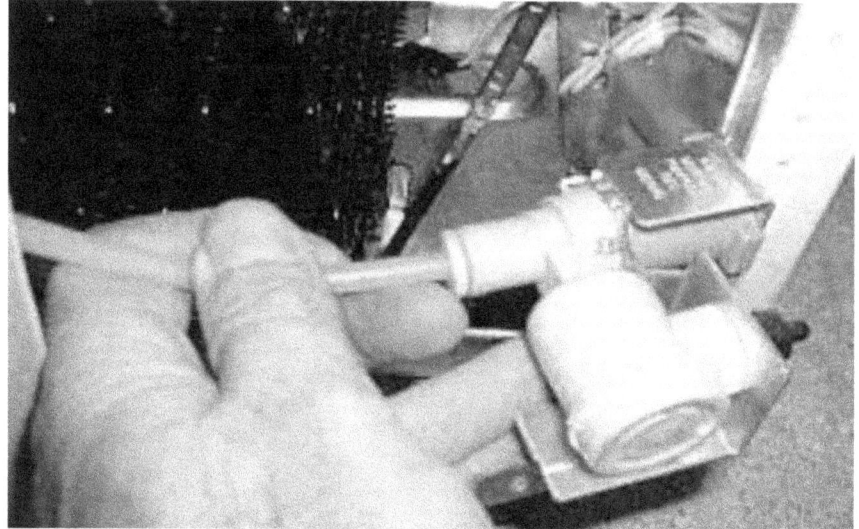

Use the manufacturer's instructions to guide you on connecting the water tubing from the water fill tube to its opening on the water valve.

## Step 7: Connect the water valve wiring

On the outer side of the refrigerator, locate the cover plate protecting the valve wiring (probably at the bottom right side). Remove the cover to expose a plastic connector containing two wires. Pull the plug-in connector out of the box and insert it onto the water valve.

Use the provided screws to mount the water valve in the refrigerator box, then reinstall the cover plate. This is an electrical connection; hence, ensure you confirm if the wires are properly connected or the ice maker will not work.

**Step 8: Finish - up**

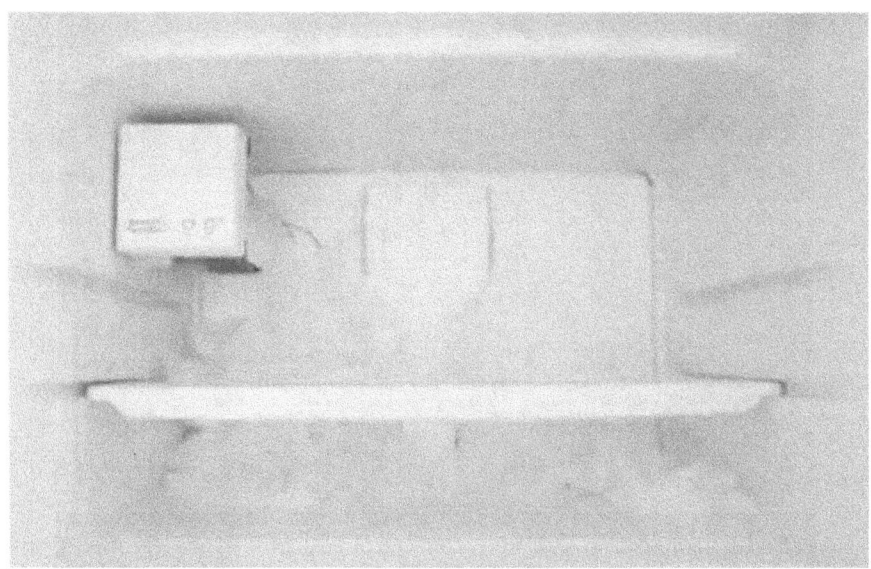

Install the connection for running water from its supply pipe to the refrigerator's inlet fitting. Again, let the manufacturer's instructions guide you on this step.

After everything is in place, plug in the refrigerator to a socket outlet and test if it is working by allowing water to flow into the icemaker. However, you will have to give the freezer some time for it to get cold before determining if you installed it well or not.

# Chapter 4: Common Wiring Problems and Their Solutions

Many wiring problems are inheritable from the previous homeowner, with many others caused by the electrician or you during wiring. Some of the common problems you are likely to face and their solutions are:

## High Electricity Bills

Outdated devices that consume a lot of power, power leakage in systems, or damaged circuits and wiring are the leading causes here.

**Solution**: Upgrade your devices to more cost-effective ones and turn off outlets or unplug devices when not in use. Also, repair or replace damaged circuits or wires.

## Overloading

Overloading occurs when you use a bulb or any other fitting of higher Watts than the fixture. This violates electricity codes and has high-risk levels. Since the bulb or fitting has high Watts, it produces high heat, which can melt wire insulations or socket outlets, leading to sparking between the wires, causing a fire.

**Solution:** Always use bulbs or fittings within the wattage range of the fixture. If the fixtures you are installing do not have watt marks, use bulbs or fittings of 60-Watts or lower.

## Tripping a Circuit Breaker

When a circuit breaker trips, it means it is doing its work: protecting you, your appliances, and your home from accidents. However, if it trips frequently, it means you are using 2 or more appliances of high Watts, e.g., hairdryer or microwave, simultaneously.

**Solution:** To solve this issue, notice the in-use appliances when the circuit breaker tripped. If possible, try to adjust its power usage to low, e.g., it is possible to reduce the power usage in hairdryers. You can also limit power usage in a circuit by using one high Watts appliance at a time.

## Electrical Surge

If you have bad power lines, poor wiring, faulty devices, or cases of lightning strikes, you are likely to experience frequent electrical surges. This electrical problem is common, and although it only lasts for a microsecond, it can damage electrical devices, reducing their life expectancy.

**Solution:** Try disconnecting any poor-quality appliance or power board connected to an outlet. This should solve your surge problem and if not, contact a licensed electrician.

**Electrical Shock**

Although mild, electrical shocks always bring about unpleasant experiences, they mostly happen when you turn on or off a faulty appliance.

**Solution:** To solve this problem, try to plug in another appliance to see if you will experience the same problem. However, this solution might not be the best, as you will be risking the occurrence of another shock. It's better to be on the safe side by talking to a licensed electrician if you experience electrical shock.

## Uncovered Junction Box

Junction boxes have many connecting wires, and if not covered well, anybody can come into contact with damaged wires. This electrical code violation can lead to many risks, especially in cases of exposed or damaged wires.

**Solution:** Always ensure all the junction boxes in your home are covered with the provided cover plates.

## Loose Wire Connection in Socket Outlets or Switches

Most screws terminals connecting wires to sockets and switches tend to get loose since they are more frequently used than other fixtures in an electrical system. This leads to crackling/buzzing or flickering sound in fixtures.

**Solution:** To solve this problem, turn off the power to the suspected light fixture, switch, or socket outlet. Remove the cover plate protecting the wires, then inspect and tighten any loose wires. Also, if wire connectors are used, check to ensure that they are secure and tight.

## 2 or More Wires Connected To a Single Screw Terminal

This common wiring problem indicates that an amateur did the work; it can be a fire hazard. It doesn't matter if the wires are of the same type, i.e., ground, neutral or live; it is a code violation to have 2 or more wires connected to the same screw terminal.

**Solution**: When you come across this kind of connection, unscrew the 2 wires on the same terminal and connect them to the terminal using a pigtail connection. Remember to use a pigtail of the same gauge as the circuit wire, e.g., use a 14 gauge pigtail in 15 amp circuit and a 12 gauge wire in 20 amperage circuit.

## Power Dips and Sags

Power dips and sags occur when a power outlet is faulty or of poor quality. This problem causes the outlet to draw a lot of power, damaging the grid or the connected appliances.

**Solution**: Call an electrician to confirm you are experiencing power dips and sags. If this is the case, replace the faulty or low-quality outlet with a new and better one.

## Backstabbing Wires

Wires connected using wire connectors at the back of those connected to the screw terminals tend to become loose. Although this is not a code violation, these wires can backstab the main connection, causing the socket or switch to stop functioning.

**Solution:** To solve this problem, confirm if the main connection is being backstabbed in the wire connectors and tighten any loose connections. Also, if there is no need for connectors, avoid using them.

Electrical problems lead to accidents or damaged appliances. The best way to avoid this is to be mindful of the signs and fix the problems immediately after noticing them. Remember that there is no guarantee that one electrical problem is related to another. Hence, always deal with each problem on its own.

# Conclusion

How does it feel to be "the guy" who does simple wiring projects inside and outside your home? It is fulfilling to learn new wiring skills during every project and keep yourself busy with something productive, isn't it?

However, it is important to keep in mind that in wiring, everything is where it is or done in a particular way for a reason. Do not take chances if you have doubts during or after any project. It is better to call a professional (even when you're halfway through projects) because a qualified electrician will help you realize where you made a mistake. Hence, you will avoid the mistake the next time, and you won't put yourself or other people at risk.